Anna Luther

Vector Field Approximation on Regular Surfaces

Anna Luther

Vector Field Approximation on Regular Surfaces

In Terms of Outer Harmonic Representations

VDM Verlag Dr. Müller

Impressum/Imprint (nur für Deutschland/ only for Germany)
Bibliografische Information der Deutschen Nationalbibliothek: Die Deutsche Nationalbibliothek verzeichnet diese Publikation in der Deutschen Nationalbibliografie; detaillierte bibliografische Daten sind im Internet über http://dnb.d-nb.de abrufbar.

Coverbild: www.ingimage.com

Verlag: VDM Verlag Dr. Müller GmbH & Co. KG
Dudweiler Landstr. 99, 66123 Saarbrücken, Deutschland
Telefon +49 681 9100-698, Telefax +49 681 9100-988
Email: info@vdm-verlag.de
Zugl.: Kaiserslautern, University, Diss., 2007

Herstellung in Deutschland:
Schaltungsdienst Lange o.H.G., Berlin
Books on Demand GmbH, Norderstedt
Reha GmbH, Saarbrücken
Amazon Distribution GmbH, Leipzig
ISBN: 978-3-8364-9667-4

Imprint (only for USA, GB)
Bibliographic information published by the Deutsche Nationalbibliothek: The Deutsche Nationalbibliothek lists this publication in the Deutsche Nationalbibliografie; detailed bibliographic data are available in the Internet at http://dnb.d-nb.de.

Cover image: www.ingimage.com

Publisher: VDM Verlag Dr. Müller GmbH & Co. KG
Dudweiler Landstr. 99, 66123 Saarbrücken, Germany
Phone +49 681 9100-698, Fax +49 681 9100-988
Email: info@vdm-publishing.com

Printed in the U.S.A.
Printed in the U.K. by (see last page)
ISBN: 978-3-8364-9667-4

Achnowledgement

First of all, I thank Prof. Dr. W. Freeden for giving me the opportunity to work on this topic and for his guidance during the development of this thesis. His valuable advice mainly contributed to the progress of this work.
Moreover, I thank all the members of the Geomathematics Group Kaiserlautern. Especially HDoz. Dr. Volker Michel, Dr. Thorsten Maier, and Dr. Carsten Mayer, for having always an open door and giving valuable comments.
I am deeply grateful to Simone Gramsch for reading the manuscript and I thank Claudia Korb for being always helpful on any concern I had.
I am grateful to Prof. Dr. habil. Gebhard Schüler and Hans Mack for their cooperation during the project work. Further, the financial support by the 'Stiftung Rheinland-Pfalz für Innovation' is gratefully acknowledged.
Finally, I am indebted to my parents Walba and Ernst Horn and especially to my husband Frank for the benefit giving me every day, all the patience and encouragement.

Contents

Chapter 1

Introduction

The intention of forestal-structure strategy and the consequential reforestation focus on the establishment of medium- and long-term ecologically robust forest stocks. The decision on positional stability of different types of trees is among other influences dependent on the modelling of data of the wind field. These observational quantities are acquired in Rheinland-Pfalz at 15 stations by the Forest Research Institute Rheinland-Pfalz ("Forschungsanstalt für Waldökologie und Forstwirtschaft (FAWF) in Rheinland-Pfalz"). For the evaluation, however, one is interested in a continuously over the surface distributed smooth representation of the wind field on the basis of the finite set of data, where smooth means that the resulting vector functions are infinitely often differentiable and that oscillations of the approximant should be avoided. Therefore in this thesis we present an approach to model the wind field by taking into account the vectorial nature of the data, thereby taking advantage of harmonic vector fields to achieve smoothness. This means that we operate on vectors instead of speed and direction values which have a scalar nature. Using harmonic vector fields to model the wind field does not include a physically relevant impact but concentrates on the creation of a smooth vector field by taking only a finite set of data into account.

In general this can be addressed as the problem of representing vector fields on regular surfaces, as e.g., the Earth's topography. For that objective we first face the problem of the exact calculation of scalar and vector outer harmonics and based on that in a second step we develop a truncated Fourier representation and a spline interpolation for restrictions of harmonic vector fields on regular surfaces. Therefore we extend the scalar approach as developed in [8, 18, 21] to the vector case.

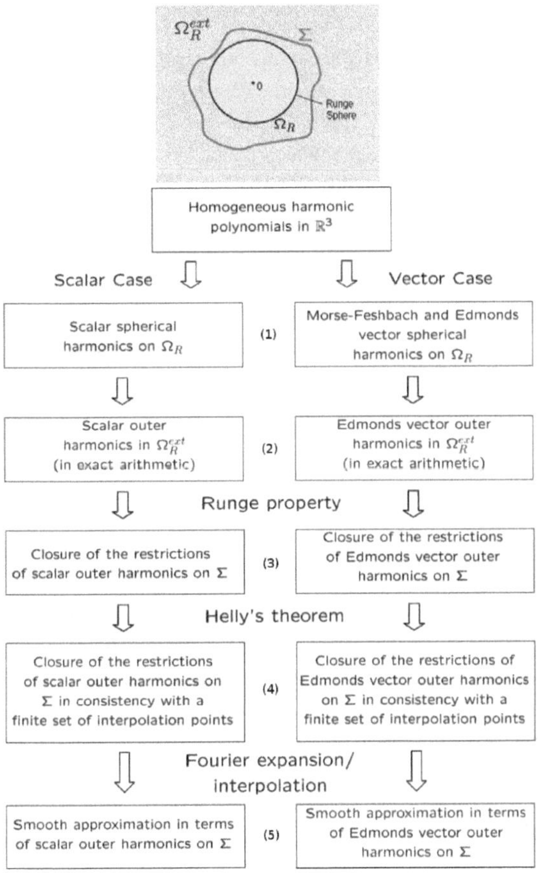

Figure 1.0.1: Geometrical concept for Ω, Ω_R^{ext} and Σ and the development steps from polynomials in $\mathbb{R}^3$ up to the approximation on Σ for the scalar and vector case in comparison.

Therefore, as presented in Figure 1.0.1 starting with the system of homogenous harmonic polynomials in $\mathbb{R}^3$ we follow the steps (1)-(5) as done in the scalar theory to develop a smooth approximation on a regular surface, denoted by Σ.
In more detail, from the homogenous harmonic polynomials which build a basis in $\mathbb{R}^3$ we derive in step (1) two kind of systems, the Morse-Feshbach and the Edmonds-system of vector spherical harmonics on a sphere Ω_R. We present an algorithm for the exact calculation of vector spherical harmonics which is applied for both systems. As in the scalar case step (2) involves the development of outer harmonics for the space outside of a sphere. In this work we use the outer harmonics which are derived from the (Edmonds-)system of vector spherical harmonics. Based on the algorithm for the exact calculation of vector spherical harmonics we provide numerical calculations for vector outer harmonics. The Runge property [35] enables us in step (3) to show that the restrictions of outer harmonics on Σ inherit the closure property. The closure property in connection with Helly's theorem [37] guarantees in step (4) the consistency for an approximate set of data resulting in step (5) in a smooth approximation on Σ by the usage of a Fourier expansion in terms of vector outer harmonics.

Our first main task focuses on the representation of an algorithm for the exact generation of scalar outer harmonics, based on the exact generation of homogeneous harmonic polynomials. For the representation of linearly independent systems of homogeneous harmonic polynomials two algorithms exclusively using integer operations are presented. The first algorithm [19] is based on the solution of an underdetermined system of linear equations, whereas the second algorithm uses a recursion relation for two-dimensional homogeneous polynomials as proposed in [20]. The exact generation of homogenous harmonic polynomials contains besides the determination of linearly independent systems also their orthonormalization. With that preparations it easy to extend the methods to the calculation of scalar spherical harmonics and scalar outer harmonics.
For the vector case we determine orthonormal systems of vector spherical harmonics in terms of cartesian coordinates. Usually (see, e.g., [6]), the numerical realization of vector spherical harmonics is based on the use of associated Legendre polynomials. However, when differentiating the associated Legendre polynomial to obtain vector spherical harmonics the problem of having singularities at the poles, arises. In this thesis we present an algorithm for constructing homogenous harmonic polynomials in cartesian coordinates with exact integer arithmetic thereby avoiding problems arising when using a local coordi-

nate system. The results are illustrated and extended to calculate vector outer harmonics which then serve as a basis for further considerations.

Equipped with the possibility to generate vector outer harmonics for any degree and order we develop Fourier series expansions for vector outer harmonics. For that purpose, we use the vector outer harmonics, introduced in [33], as basis functions for the outer space of a sphere. The theoretical backbone is provided by the closure and completeness of restrictions of vector outer harmonics on regular surfaces. In addition to the property of closure, the interpolation property for a finite set of approximation points can be guaranteed by Helly's theorem [37]. The procedure as described in [18, 21] is then extended to the vector case.

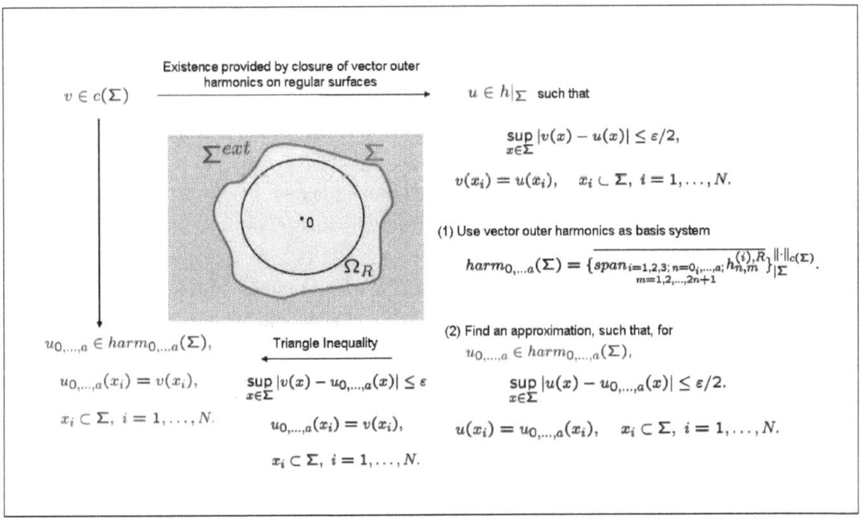

Figure 1.0.2: Approximation of a continuous vector function.

Figure 1.0.2 illustrates the construction principles for the approximation of continuous functions which are described in more detail in the following. Let Σ be a regular surface and denote the interior of this surface by Σ^{int}. The approximate function is assumed to satisfy the Laplace equation outside an arbitrarily given sphere Ω_R inside the inner space Σ^{int}. The closure and completeness of vector outer harmonics in connection with Helly's theorem shows that, corresponding to the continuous vector function v on Σ, there exists a member u of a reference space $h|_\Sigma$ in an $(\varepsilon/2)-$neighborhood, such that the values of u are consistent with the function values of the continuous vector function v on Σ for the known finite set of discrete points. Moreover, this function u of class $h|_\Sigma$ may be considered to be in $(\varepsilon/2)-$accuracy to a member $u_{0,\ldots,a}$ of a set of vector outer harmonics up to degree a, restricted to Σ, $h_{0,\ldots,a}|_\Sigma$, which can be supposed to be consistent with the known function values as well. Thus, to any continuous vector function v on a regular surface Σ, there exists in ε-accuracy a bandlimited vector function $u_{0,\ldots,a} \in h_{0,\ldots,a}|_\Sigma$ such that this bandlimited vector function coincides at all given points with the function values of the original continuous vector function on the regular surface Σ.

The objective of our work, is to show that the approximation can be established in a constructive way as an (orthogonal) Fourier series for vector outer harmonics. Our interest lies in a Fourier approximation of a function $u_{0,\ldots,a}$ of class $h_{0,\ldots,a}|_\Sigma$ from discretely given vector function values on Σ. The method is a generalization of the scalar Fourier variant (second variant of [21]) due to Freeden and Schneider. First, we introduce a reference space and give the representation of a reproducing kernel, constituted from vector outer harmonics. Then we introduce a new class of approximate formulae involving vector outer harmonics.

Next, we are concerned with the approximation of continuous vector functions on regular surfaces corresponding to scattered vector function values on the finite set of discrete points (on the regular surface). For the case having only a discrete set of vector data we discuss the spline interpolation problem for smooth vector functions on regular surfaces. Taking into account the considerations developed for the Fourier representation of vector outer harmonics we deduce that by observing restrictions of continuous vector functions on Σ there is a possibility to find in $\varepsilon-$accuracy vector outer harmonics such that the interpolation property is assured.

The outline of this thesis is as follows.

The second chapter provides the basic notation and defines the spaces and differential operators for a spherical set up. In this chapter we also introduce Legendre polynomials and describe what we mean when we designate regular surfaces.

Chapter 3 gives an overview on spherical polynomials. First the definition of homogeneous harmonic polynomials and their addition theorem are presented. Then scalar spherical harmonics and two kinds of vector spherical harmonics (the (Morse-Feshbach-) system and the (Edmonds-)vector spherical harmonics) relating to different properties when regarding the Laplace equation are introduced, as, e.g., in [14]. The (Edmonds-)system of vector spherical harmonics has the property to be a set of eigenfunctions to the Beltrami operator. Thus we are able to define a set of vector functions which fulfill the Laplace equation in the outer space of a sphere with radius R.

In Chapter 4 we first introduce scalar outer harmonics and then develop vector outer harmonics in such a way, that the Laplace equation is fulfilled componentwise. The closure property of the vector outer harmonics system is shown which allows to use these functions as a basis for the approximation of continuous vector fields on regular surfaces. In Section 4.3 we present two ways for the exact calculation of homogeneous harmonic polynomials. First by solving underdetermined systems and then via recursion relations, followed by the calculation of scalar spherical harmonics and scalar outer harmonics and the corresponding illustrations. Section 4.4 provides the exact generation of vector spherical harmonics and vector outer harmonics and provides also illustrations of these functions.

In Chapter 5 we introduce first the reference space in which a reproducing kernel structure can be set up and use then this space for the Fourier representation of vector outer harmonics (similar to the scalar case, as proposed in [10, 18, 21]). Here, we show that we are able to present a fully discrete Fourier approximation for a vector function on a regular surface. Section 5.3 deals with the problem to find the smoothest vector field for a continuous function on a regular surface from given function values. The result is presented in a spline interpolation procedure taking into account the reproducing kernel structure of the used reference space. This chapter closes with numerical examples for the Fourier approximation of vector functions on regular surface for discretely given wind field measurements over Palatinate. Thus the last numerical example focuses without any further physical information as, e.g., the pressure, on a smooth modelling of the wind field over the given topography.

Chapter 2

Preliminaries

This chapter introduces some of the basic mathematical tools to build a basis on which the following chapters rely.
In this work we operate with two different coordinate systems, therefore the reader obtains an overview about the notation for the Euclidean space $\mathbb{R}^3$ and the corresponding cartesian coordinate system as well as the spherical coordinates and the related spherical nomenclature. We define then the relevant scalar and vector function spaces and introduce differential operators, in particular operators which are used to generate vector fields. A short description of a special system of polynomials, called Legendre polynomials, is given followed by geometrical assumptions containing information about regular surfaces.
The preliminaries are mainly due to [11, 14, 18].

Euclidean Space $\mathbb{R}^3$

First we give the basic notation for the Euclidean space and set up the notation for spherical problems. The three-dimensional Euclidean space is denoted by $\mathbb{R}^3$. With the canonical orthonormal basis

$$\varepsilon^1 = \begin{pmatrix} 1 \\ 0 \\ 0 \end{pmatrix}, \ \varepsilon^2 = \begin{pmatrix} 0 \\ 1 \\ 0 \end{pmatrix}, \ \varepsilon^3 = \begin{pmatrix} 0 \\ 0 \\ 1 \end{pmatrix}, \tag{2.0.1}$$

any vector $x \in \mathbb{R}^3$ can uniquely be written as $x = x_1\varepsilon^1 + x_2\varepsilon^2 + x_3\varepsilon^3$. Thus elements x, $y \in \mathbb{R}^3$ can be represented by using their components with respect to the Cartesian basis, i.e., $x = (x_1, x_2, x_3)^T$, $y = (y_1, y_2, y_3)^T$. Basic operations for the Euclidean basis are given by the scalar product,

$$x \cdot y = x_1 y_1 + x_2 y_2 + x_3 y_3,$$

the norm

$$|x| = \sqrt{x^2} = \sqrt{x_1^2 + x_2^2 + x_3^2},$$

the vector product

$$x \wedge y = (x_2 y_3 - x_3 y_2, x_3 y_1 - x_1 y_3, x_1 y_2 - x_2 y_1),$$

and the tensor product

$$x \otimes y = \begin{pmatrix} x_1 y_1 & x_1 y_2 & x_1 y_3 \\ x_2 y_1 & x_2 y_2 & x_1 y_3 \\ x_3 y_1 & x_3 y_2 & x_3 y_3 \end{pmatrix}.$$

For vectors $x, y, w, z \in \mathbb{R}^3$ it is easy to verify that

$$(x \otimes y)(w \otimes z) = (y \cdot w)(x \otimes z)$$

and

$$(x \otimes y)z = (y \cdot z)x. \tag{2.0.2}$$

Furthermore, let us introduce some notation needed for the representation of polynomial functions. Let $\alpha_i \in \mathbb{N}_0$ for $i = 1, 2, 3$. We denote $\alpha = (\alpha_1, \alpha_2, \alpha_3)^T$, with $\alpha_1, \alpha_2, \alpha_3 \geq 0$ as the multiindex. The factorial of the multiindex is given by $\alpha! = \alpha_1!\,\alpha_2!\,\alpha_3!$ and the degree of the multiindex is defined by $[\alpha] = \alpha_1 + \alpha_2 + \alpha_3$. We write $x^\alpha = x_1^{\alpha_1} x_2^{\alpha_2} x_3^{\alpha_3}$ for abbreviation. With the binomial theorem we obtain the following two relations

$$(x_1 + x_2 + x_3)^n = \sum_{[\alpha]=n} \frac{n!}{\alpha_1!\alpha_2!\alpha_3!} x_1^{\alpha^1} \cdot x_2^{\alpha^2} \cdot x_3^{\alpha^3} = \sum_{[\alpha]=n} \frac{n!}{\alpha!} x^\alpha,$$

$$(x \cdot y)^n = (x_1 y_1 + x_2 y_2 + x_3 y_3)^n = \sum_{[\alpha]=n} \frac{n!}{\alpha!} x^\alpha y^\alpha.$$

Spherical Nomenclature

The sphere $\Omega_R \in \mathbb{R}^3$ with radius $R > 0$, $R \in \mathbb{R}$, is defined by

$$\Omega_R = \{x \in \mathbb{R}^3 | \ |x| = R\}.$$

The unit sphere is denoted by Ω, i.e., $\Omega_1 = \Omega$. For the total surface of Ω we have $\|\Omega\| = \int_\Omega d\omega = 4\pi$, where $d\omega$ denotes the surface element.
Any $x \in \mathbb{R}^3 \setminus \{0\}$ can be written as $x = r\xi$ with $r = |x|$ and $\xi = (\xi_1, \xi_2, \xi_3)^T \in \Omega$, i.e., x is separated into its length r and its direction ξ. We can write

$$x = \begin{pmatrix} r\sqrt{1-t^2}\cos\varphi \\ r\sqrt{1-t^2}\sin\varphi \\ rt \end{pmatrix}, \quad \varphi \in [0, 2\pi), \ \ t \in [-1, 1], \ \ r = |x|, \tag{2.0.3}$$

where t is the polar distance and φ is the spherical longitude.

Function Spaces

We use the following general scheme of notation throughout this work:

capital letters F, G	:	scalar functions
lower-case letters f, g	:	vector fields
boldface lower-case letters **f**,**g**	:	tensor fields of second rank

$\mathcal{L}^p(\Omega)$ is the class of (scalar) functions $F : \Omega \to \mathbb{R}$ that are measurable with

$$\|F\|_{\mathcal{L}^p(\Omega)} = \left(\int_\Omega |F(\eta)|^p d\omega(\eta)\right)^{1/p} < \infty, \qquad 1 \le p < \infty.$$

$\mathcal{L}^p(\Omega)$ admits the inclusion

$$\mathcal{L}^p(\Omega) \subset \mathcal{L}^q(\Omega), \qquad 1 \le q < p.$$

For $p = 2$ and with respect to the inner product

$$(F, G)_{\mathcal{L}^2(\Omega)} = \int_\Omega F(\eta)G(\eta)d\omega(\eta),$$

$\mathcal{L}^2(\Omega)$ is a Hilbert space. It is the space of all measurable and square-integrable scalar functions on the sphere.
The Banach space $\mathcal{C}(\Omega)$ is given by

$$\mathcal{C}(\Omega) = \{F : \Omega \to \mathbb{R} \,|\, F \text{ continuous on } \Omega\},$$

equipped with the norm

$$\|F\|_{\mathcal{C}(\Omega)} = \sup_{\xi\in\Omega} |F(\xi)|.$$

$C^{(k)}(\Omega)$ denotes the class of k times continuously differentiable scalar functions $F : \Omega \to \mathbb{R}$. As is well known, $\mathcal{L}^2(\Omega)$ is the completion of $\mathcal{C}(\Omega)$ with respect to $\|\cdot\|_{\mathcal{L}^2(\Omega)}$, that is

$$\mathcal{L}^2(\Omega) = \overline{\mathcal{C}(\Omega)}^{\|\cdot\|_{\mathcal{L}^2(\Omega)}}. \tag{2.0.4}$$

There is the following norm estimate between the $\mathcal{C}-$ and the $\mathcal{L}^2-$norm:

$$\|F\|_{\mathcal{L}^2(\Omega)} = \left(\int_\Omega |F(\eta)|^2 d\omega(\eta)\right)^{1/2} \le \sqrt{4\pi}\|F\|_{\mathcal{C}(\Omega)}, \quad F \in \mathcal{C}(\Omega).$$

Let $\xi \in \Omega$ be fixed, then for every $\eta \in \Omega$ the product of ξ and η satisfies $-1 \le \xi\cdot\eta \le 1$. Hence for every function $G \in \mathcal{L}^2[-1,1]$ we can define a ξ-zonal function $G_\xi \in \mathcal{L}^2(\Omega)$ by

$$\eta \mapsto G_\xi(\eta) = G(\xi\cdot\eta). \tag{2.0.5}$$

The value of $G_\xi(\eta)$ is constant on the set

$$M(\xi, t) = \{\eta \in \Omega \,|\, \xi\cdot\eta = t\}, \quad t \in [-1,1],$$

it depends only on the polar distance t between $\xi \in \Omega$ and $\eta \in \Omega$. The set of all ξ-zonal functions is isomorphic to the set of functions $G : [-1,1] \to \mathbb{R}$. Thus we can interpret $\mathcal{C}[-1,1]$ and $\mathcal{L}^p[-1,1]$ by norms defined as subspaces of $\mathcal{C}(\Omega)$ and $\mathcal{L}^p(\Omega)$, respectively.

$$\|G\|_{\mathcal{L}^p[-1,1]} = \|G(\varepsilon^3\cdot)\|_{\mathcal{L}^p(\Omega)} = \left(2\pi\int_{-1}^{1} |G(s)|^p ds\right)^{1/p},$$

$$\|G\|_{\mathcal{C}[-1,1]} = \|G(\varepsilon^3\cdot)\|_{\mathcal{C}(\Omega)} = \sup_{\eta\in\Omega} |G(\varepsilon^3\cdot\eta)| = \sup_{s\in[-1,1]} |G(s)|.$$

In the following we are interested in spherical vector fields. Using the canonical orthonormal basis of $\mathbb{R}^3$, as given in (2.0.1), every vector field $f : \Omega \to \mathbb{R}^3$ can be represented by

$$f(\xi) = \sum_{i=1}^{3} F_i(\xi)\varepsilon^i, \quad \xi \in \Omega, \tag{2.0.6}$$

where F_i are the component functions of f given by the projections onto the basis vectors, i.e., $F_i(\xi) = f(\xi) \cdot \varepsilon^i$, $i = 1, 2, 3$, $\xi \in \Omega$.
With $l^2(\Omega)$ we denote the class of all square-integrable vector fields on Ω. Equipped with the inner product

$$(f, g)_{l^2(\Omega)} = \int_\Omega f(\eta) \cdot g(\eta) d\omega(\eta), \quad f, g \in l^2(\Omega),$$

and the norm

$$\|f\|_{l^2(\Omega)} = \left(\int_\Omega |f(\eta)|^2 d\omega(\eta) \right)^{1/2}, \quad f \in l^2(\Omega),$$

$l^2(\Omega)$ is a Hilbert space. The q-times continuous differentiability of a vector field f is given if the component functions F_i are q-times continuously differentiable.

The space $c^{(q)}(\Omega)$, $0 \leq q \leq \infty$, consists of all q-times continuously differentiable vector fields on Ω. Endowed with the norm

$$\|f\|_{c(\Omega)} = \sup_{\xi \in \Omega} |f(\xi)|, \quad f \in c(\Omega),$$

the space $c(\Omega)$ is complete. Analogously to the scalar case, $l^2(\Omega)$ is the completion of $c(\Omega)$ with respect to the $l^2(\Omega)$-norm, i.e.,

$$l^2(\Omega) = \overline{c(\Omega)}^{\|\cdot\|_{l^2(\Omega)}}.$$

For $f \in c(\Omega)$ we have the norm estimate

$$\|f\|_{l^2(\Omega)} \leq \sqrt{4\pi} \|f\|_{c(\Omega)}.$$

Introducing the projection operators p_{nor} and p_{tan} we are able to decompose a vector field into its normal and tangential part, respectively. In more detail, for $\xi \in \Omega$, we define the projection operators by

$$\begin{aligned} p_{nor} f(\xi) &= (f(\xi) \cdot \xi), \\ p_{tan} f(\xi) &= f(\xi) - p_{nor} f(\xi), \end{aligned} \tag{2.0.7}$$

acting on continuous vector fields on the sphere. Obviously we can write

$$\begin{aligned} c_{nor}(\Omega) &= \{ f \in c(\Omega) | \, f = p_{nor} f \}, \\ c_{tan}(\Omega) &= \{ f \in c(\Omega) | \, f = p_{tan} f \}, \end{aligned} \tag{2.0.8}$$

and combine them to

$$c(\Omega) = c_{nor}(\Omega) \oplus c_{tan}(\Omega). \tag{2.0.9}$$

With (2.0.7) we are also able to introduce projection operators for vector fields $f \in l^2(\Omega)$, leading to a decomposition of the Hilbert space $l^2(\Omega)$ into two orthogonal parts

$$\begin{aligned} l^2_{nor}(\Omega) &= \{f \in l^2(\Omega) | f = p_{nor} f\}, \\ l^2_{tan}(\Omega) &= \{f \in l^2(\Omega) | f = p_{tan} f\}. \end{aligned} \tag{2.0.10}$$

This yields the orthogonal decomposition of $l^2(\Omega)$:

$$l^2(\Omega) = l^2_{nor}(\Omega) \oplus l^2_{tan}(\Omega). \tag{2.0.11}$$

Differential Operators

In Table 2.1 a number of differential operators are listed (see for more details [14]).

Table 2.1: Differential operators

Symbol	**Differential Operator**
∇_x	Gradient at x
$\Delta_x = \nabla_x \cdot \nabla_x$	Laplace operator at x
∇^*_ξ	Surface gradient on the unit sphere Ω at ξ
$L^*_\xi = \xi \wedge \nabla^*_\xi$	Surface curl gradient on the unit sphere Ω at ξ
$\Delta^*_\xi = \nabla^*_\xi \cdot \nabla^*_\xi = L^*_\xi \cdot L^*_\xi$	Beltrami operator on the unit sphere Ω at ξ
$\nabla^* \cdot$	Surface divergence on the unit sphere Ω at ξ
$L^* \cdot$	Surface curl on the unit sphere Ω at ξ

For the convenience of the reader these operators will be discussed in the particular system of polar coordinates for $x = r\xi$, $\xi \in \Omega$, where the split into a radial and angular part is provided.

As usual, the gradient operator can be written by

$$\nabla_x = \left(\frac{\partial}{\partial x_1}, \frac{\partial}{\partial x_2}, \frac{\partial}{\partial x_3} \right)^T. \tag{2.0.12}$$

For the purpose of expressing polynomials in terms of multiindices we introduce

$$(\nabla_x)^\alpha = \left(\frac{\partial}{\partial x_1}\right)^{\alpha_1}\left(\frac{\partial}{\partial x_2}\right)^{\alpha_2}\left(\frac{\partial}{\partial x_3}\right)^{\alpha_3} = \frac{\partial^{[\alpha]}}{(\partial x_1)^{\alpha_1}(\partial x_2)^{\alpha_2}(\partial x_3)^{\alpha_3}}.$$

Obviously, we have

$$(\nabla_x)^\alpha x^\beta = \begin{cases} 0, & \text{for } \alpha \neq \beta \text{ and } [\alpha] = [\beta], \\ \alpha!, & \text{for } \alpha = \beta. \end{cases} \tag{2.0.13}$$

The Laplacian

$$\Delta_x = \left(\frac{\partial}{\partial x_1}\right)^2 + \left(\frac{\partial}{\partial x_2}\right)^2 + \left(\frac{\partial}{\partial x_3}\right)^2 \tag{2.0.14}$$

can also be formally written as $\Delta_x = \nabla_x \cdot \nabla_x$. The operator $L_x = x \wedge \nabla_x$ is said to be the curl gradient.

By setting $r = 1$ in (2.0.3) we obtain a local coordinate system on the unit sphere Ω. In doing so we obtain basis vectors

$$\varepsilon^r(\varphi, t) = \left(\frac{\partial x(r,\varphi,t)}{\partial r}\right)\Big|_{r=1} \left|\left(\frac{\partial x(r,\varphi,t)}{\partial r}\right)\Big|_{r=1}\right|^{-1},$$

$$\varepsilon^\varphi(\varphi, t) = \left(\frac{\partial x(1,\varphi,t)}{\partial \varphi}\right) \left|\left(\frac{\partial x(1,\varphi,t)}{\partial \varphi}\right)\right|^{-1},$$

$$\varepsilon^t(\varphi, t) = \left(\frac{\partial x(1,\varphi,t)}{\partial t}\right) \left|\left(\frac{\partial x(1,\varphi,t)}{\partial t}\right)\right|^{-1},$$

which build a moving orthonormal triad on the unit sphere Ω. Carrying out the calculations yields

$$\varepsilon^r(\varphi, t) = \begin{pmatrix} \sqrt{1-t^2}\cos(\varphi) \\ \sqrt{1-t^2}\sin(\varphi) \\ t \end{pmatrix}, \quad \varepsilon^\varphi(\varphi, t) = \begin{pmatrix} -\sin(\varphi) \\ \cos(\varphi) \\ 0 \end{pmatrix}, \quad \varepsilon^t(\varphi, t) = \begin{pmatrix} -t\cos(\varphi) \\ -t\sin(\varphi) \\ \sqrt{1-t^2} \end{pmatrix}.$$

Notice that

$$\varepsilon^r(\varphi, t) \wedge \varepsilon^\varphi(\varphi, t) = \varepsilon^t(\varphi, t).$$

The canonical basis vectors of $\mathbb{R}^3$ can be expressed in terms of ε^r, ε^φ, ε^t in the following way:

$$\begin{aligned}\varepsilon^1 &= \sqrt{1-t^2}\cos(\varphi)\varepsilon^r(\varphi,t) - \sin(\varphi)\varepsilon^\varphi(\varphi,t) - t\cos(\varphi)\varepsilon^t(\varphi,t),\\ \varepsilon^2 &= \sqrt{1-t^2}\sin(\varphi)\varepsilon^r(\varphi,t) - \cos(\varphi)\varepsilon^\varphi(\varphi,t) - t\sin(\varphi)\varepsilon^t(\varphi,t),\\ \varepsilon^3 &= t\varepsilon^r(\varphi,t) + \sqrt{1-t^2}\varepsilon^t(\varphi,t).\end{aligned} \tag{2.0.15}$$

Using the polar coordinates in (2.0.12) and (2.0.14) we can separate the gradient and the Laplace operator into a purely tangential and a purely radial part by

$$\nabla = \varepsilon^r \frac{\partial}{\partial r} + \frac{1}{r}\nabla^*, \tag{2.0.16}$$

where ∇^*, the surface gradient on Ω, is the angular part of ∇ given by

$$\nabla^* = \varepsilon^\varphi \frac{1}{\sqrt{1-t^2}}\frac{\partial}{\partial\varphi} + \varepsilon^t\sqrt{1-t^2}\frac{\partial}{\partial t}. \tag{2.0.17}$$

For the Laplace operator we obtain

$$\Delta = \left(\frac{\partial}{\partial r}\right)^2 + \frac{2}{r}\frac{\partial}{\partial r} + \frac{1}{r^2}\Delta^*, \tag{2.0.18}$$

where Δ^*, the Beltrami operator on the unit sphere, is the angular part of the Laplace operator

$$\Delta^* = \frac{\partial}{\partial t}(1-t^2)\frac{\partial}{\partial t} + \frac{1}{1-t^2}\left(\frac{\partial}{\partial\varphi}\right)^2. \tag{2.0.19}$$

Later on, working with vector spherical harmonics, we will need the so-called surface curl gradient:

$$L^*_\xi = \xi \wedge \nabla^*_\xi, \quad \xi \in \Omega. \tag{2.0.20}$$

Applying L^* to $F \in \mathcal{C}^{(1)}(\Omega)$ we obtain

$$L^*_\xi F(\xi) = \xi \wedge \nabla^*_\xi F(\xi),$$

which is a tangential vector field, perpendicular to the vector field $\nabla^*_\xi F(\xi)$. Using local coordinates we are able to write

$$L^* = -\varepsilon^\varphi\sqrt{1-t^2}\frac{\partial}{\partial t} + \varepsilon^t \frac{1}{\sqrt{1-t^2}}\frac{\partial}{\partial\varphi}. \tag{2.0.21}$$

For a fixed $\eta \in \Omega$ the following relations hold:

$$\begin{aligned}\nabla^*_\xi(\xi\cdot\eta) &= \eta-(\xi\cdot\eta)\xi,\\ L^*_\xi(\xi\cdot\eta) &= \xi\wedge\eta,\\ \Delta^*_\xi(\xi\cdot\eta) &= -2(\xi\cdot\eta).\end{aligned}$$

Applying the operators on radial basis functions as introduced by (2.0.5) leads to the following results:
Let $F \in \mathcal{C}^{(1)}[-1,1]$, then for $\xi,\eta \in \Omega$

$$\begin{aligned}\nabla^*_\xi F(\xi\cdot\eta) &= F'(\xi\cdot\eta)(\eta-(\xi\cdot\eta)\xi),\\ L^*_\xi F(\xi\cdot\eta) &= F'(\xi\cdot\eta)(\xi\wedge\eta),\end{aligned}$$

whereas for $F \in \mathcal{C}^{(2)}[-1,1]$, we get

$$\Delta^*_\xi F(\xi\cdot\eta) = -2(\xi\cdot\eta)F'(\xi\cdot\eta)+(1-(\xi\cdot\eta)^2)F''(\xi\cdot\eta),\quad \xi,\eta\in\Omega. \tag{2.0.22}$$

The (formal) scalar product of the operators ∇^* and L^* with a tangential vector field $f \in c^{(1)}_{tan}(\Omega)$, $f(\xi)=\sum_{i=1}^3 F_i(\xi)\varepsilon^i$, yields the surface divergence $\nabla^*_\xi\cdot$ defined by

$$\nabla^*_\xi\cdot f(\xi) = \sum_{i=1}^3(\nabla^*_\xi F_i(\xi))\cdot\varepsilon^i \tag{2.0.23}$$

and the surface curl $L^*_\xi\cdot$ given by

$$L^*_\xi\cdot f(\xi) = \sum_{i=1}^3(L^*_\xi F_i(\xi))\cdot\varepsilon^i. \tag{2.0.24}$$

Remark 2.0.1. *The motivation for the definition of the operators ∇^* and L^* is the fact that any $f \in c^{(1)}(\Omega)$ can be decomposed by the Helmholtz Theorem (see [14]) as follows*

$$f(\xi) = \xi F_1(\xi)+\nabla^*_\xi F_2(\xi)+L^*_\xi F_3(\xi), \tag{2.0.25}$$

with functions $F_i : \Omega \to \mathbb{R}$, $i=1,2,3$, satisfying

$$\int_\Omega F_2(\eta)d\omega(\eta) = \int_\Omega F_3(\eta)d\omega(\eta) = 0$$

and

$$p_{tan}f(\xi) = \nabla^*_\xi F_2(\xi)+L^*_\xi F_3(\xi),\ \xi\in\Omega.$$

The last part of this section introduces operators o^i, $i = 1, 2, 3$. Throughout the work we will use the following notation

$$0_i = \begin{cases} 0, & \text{if } i = 1, \\ 1, & \text{if } i = 2, 3. \end{cases} \tag{2.0.26}$$

Regarding notation (2.0.26) we will use the following abbreviation $\mathbb{N}_{0_i} = \{n \in \mathbb{N} \,|\, n \geq 0_i\}$. For $F \in C^{(1)}(\Omega)$ the operators $o^{(i)} : C^{(0_i)}(\Omega) \to c(\Omega)$, $i = 1, 2, 3$, are defined by

$$\begin{aligned} o_\xi^{(1)} F(\xi) &= \xi F(\xi), \\ o_\xi^{(2)} F(\xi) &= \nabla_\xi^* F(\xi), \\ o_\xi^{(3)} F(\xi) &= L_\xi^* F(\xi). \end{aligned} \tag{2.0.27}$$

Hence, $o^{(1)}F$ is a normal field, whereas $o_\xi^{(2)}F$ and $o_\xi^{(3)}F$ are tangential fields. Furthermore, $o_\xi^2 F(\xi)$ is curl-free, whereas $o_\xi^3 F(\xi)$ is divergence-free, as $\nabla_\xi^* F(\xi)$ is a gradient- and $L_\xi^* F(\xi)$ is a curl-field. Additionally, we see that

$$o_\xi^i F(\xi) \cdot o_\xi^j F(\xi) = 0, \quad \text{for all } i \neq j, \quad i, j \in 1, 2, 3, \quad \xi \in \Omega. \tag{2.0.28}$$

Let now $f : \Omega \to \mathbb{R}^3$ be a continuously differentiable vector field. We separate f into a normal and a tangential component

$$f = f_{nor} + f_{tan}.$$

Then there exist scalar-valued functions $F^{(i)} : \Omega \to \mathbb{R}$, $i = 1, 2, 3$ such that

$$\begin{aligned} f_{nor}(\xi) &= o_\xi^{(1)} F^{(1)}(\xi), \quad \xi \in \Omega, \\ f_{tan}(\xi) &= o_\xi^{(2)} F^{(2)}(\xi) + o_\xi^{(3)} F^{(3)}(\xi), \quad \xi \in \Omega, \end{aligned} \tag{2.0.29}$$

Notice that $F^{(i)}$, $i = 2, 3$, have to be twice continuously differentiable functions.

The functions $F^{(i)}$, $i = 1, 2, 3$, are given by

$$\begin{aligned} F^{(1)}(\xi) &= \xi \cdot f(\xi), \quad \xi \in \Omega, \\ F^{(2)}(\xi) &= -\int_\Omega G(\Delta^*; \xi; \eta) \nabla_\xi^* \cdot f(\eta) d\omega(\eta), \quad \xi \in \Omega, \\ F^{(3)}(\xi) &= -\int_\Omega G(\Delta^*; \xi; \eta) L_\xi^* \cdot f(\eta) d\omega(\eta), \quad \xi \in \Omega, \end{aligned}$$

where $G(\Delta^*;\xi;\eta)$ is Green's function with respect to the Beltrami operator Δ^*, given explicitly (see [7]) by

$$G(\Delta^*;\xi;\eta) = \frac{1}{4\pi}\ln(1-\xi\cdot\eta) + \frac{1}{4\pi} - \frac{1}{4\pi}\ln 2, \quad \xi,\eta\in\Omega.$$

Later we will use the adjoint operators of $o^{(i)}$, i.e., those operators $O^{(i)}$ for which holds

$$(o^{(i)}F, g)_{l^2(\Omega)} = (F, O^{(i)}g)_{\mathcal{L}^2(\Omega)},$$

whenever F, g are continuously differentiable functions. With the projections given by (2.0.7) these operators can be represented as follows:

$$\begin{aligned} O^{(1)}_\xi g(\xi) &= \xi\cdot p_{nor}f(\xi), \quad \xi\in\Omega,\\ O^{(1)}_\xi g(\xi) &= -\nabla^*_\xi\cdot p_{tan}f(\xi), \quad \xi\in\Omega,\\ O^{(1)}_\xi g(\xi) &= -L^*_\xi\cdot p_{tan}f(\xi), \quad \xi\in\Omega. \end{aligned} \tag{2.0.30}$$

These definitions lead to the following results: Let $F\in\mathcal{C}^{(2)}(\Omega)$, then the following relations are valid.

1. For $i\neq j$ and $i,j\in\{1,2,3\}$ the equation $O^i_\xi o^j_\xi F(\xi) = 0$ holds,
2. for $i\in\{1,2,3\}$,
$$O^i_\xi o^i_\xi F(\xi) = \begin{cases} F(\xi), & \text{for } i=1,\\ -\Delta^*_\xi F(\xi), & \text{for } i=2,3. \end{cases}$$

Legendre Polynomials

Legendre polynomials form an orthogonal basis set in $\mathcal{L}^2[-1,1]$. In the following we introduce the Legendre polynomials as done in [14].
A function

$$P_n : [-1,1]\to\mathbb{R},\ n\in\mathbb{N}_0, \tag{2.0.31}$$

satisfying the following properties

1. P_n is a polynomial of degree n,
2. $\int_{-1}^{1} P_n(t)P_l(t)dt = 0$, for $n\neq l$ (orthogonality),

3. $P_n(1) = 1$,

is called Legendre polynomial.

Legendre polynomials can uniquely be defined as eigenfunctions of the Legendre operator given by

$$L_t = \frac{d}{dt}(1-t^2)\frac{d}{dt}, \quad t \in [-1,1],$$

which is a part of the Beltrami operator as given by (2.0.19).

Definition 2.0.1. *The Legendre polynomials $P_n : [-1,1] \to \mathbb{R}$ of degree n are uniquely defined as the twice differentiable eigenfunctions of the Legendre operator corresponding to the eigenvalues -n(n+1), such that*

$$(L_t + n(n+1))P_n(t) = 0,$$

with P_n satisfying the additional condition $P_n(1) = 1$.

The Legendre polynomials form an orthogonal set with respect to the $\mathcal{L}^2[-1,1]$ - inner product, i.e.,

$$\int_{-1}^{1} P_n(x)P_m(x)dx = \frac{2}{2n+1}\,\delta_{nm}, \quad n,m \in \mathbb{N}_0,$$

with δ_{nm} being the Kronecker symbol.

Theorem 2.0.1. *Let the Legendre polynomials be given by Definition 2.0.1 and $n \in \mathbb{N}_0$. Then the following statements hold true.*

1. *The set $\{P_n\}_{n\in\mathbb{N}}$ is complete in $\mathcal{L}^2[-1,1]$ with respect to $\|\cdot\|_{\mathcal{L}^2[-1,1]}$ and closed within $\mathcal{C}[-1,1]$ with respect to $\|\cdot\|_{\mathcal{C}[-1,1]}$,*

2. *$P_n(\varepsilon^3\cdot)$ as in (2.0.5) is the only polynomial of degree n, that is invariant with respect to orthogonal transformations $\mathbf{t}$ with $\mathbf{t}\varepsilon^3 = \varepsilon^3$ (i.e., that leave the ε^3-axis fixed).*

Let F be any function of class $\mathcal{L}^2[-1,1]$, then the representation by its Legendre series is given via

$$F = \sum_{n=0}^{\infty} \frac{2n+1}{4\pi} F^\wedge(n) P_n, \tag{2.0.32}$$

with the Legendre coefficients $F^\wedge(n)$ given by

$$F^\wedge(n) = (F, P_n)_{\mathcal{L}^2[-1,1]} = \int_{-1}^{1} F(x)P_n(x)dx, \quad n \in \mathbb{N}_0.$$

The equality (2.0.32) is understood in the $\mathcal{L}^2[-1,1]$-sense, i.e.,

$$\lim_{N\to\infty} \left\| F - \sum_{n=0}^{N} \frac{2n+1}{4\pi} F^\wedge(n)P_n \right\|_{\mathcal{L}^2[-1,1]} = 0.$$

For a more detailed description about Legendre polynomials we refer to [14].

Geometrical Assumptions

For our purposes concerning the approximation of functions given on discrete points on a surface we need to explain which kind of surfaces are approved. This will be done as proposed, e.g., in [8, 18, 21].

Definition 2.0.2. *A surface $\Sigma \subset \mathbb{R}^3$ is said to be regular, if it satisfies the following properties (see Figure 2.0.1):*

1. *Σ divides the three-dimensional space $\mathbb{R}^3$ into the (open) bounded region Σ^{int} (inner space) and the (open) unbounded region Σ^{ext} (outer space) defined by $\Sigma^{ext} = \mathbb{R}^3 \setminus \overline{\Sigma^{int}}$.*
2. *Σ is a closed and compact surface with no double points.*
3. *The origin is contained in Σ^{int}.*
4. *Σ has a continuously differentiable unit normal field ν pointing into the outer space Σ^{ext}.*

Georelevant regular surfaces Σ are, for example, spheres, ellipsoids, spheroids, the geoid, the (regular) Earth's surface, etc.
As usual, Ω_R^{int} and Ω_R^{ext} denote the inner and the outer space of the sphere Ω_R around the origin with radius R. Σ_{inf}^{int}, Σ_{sup}^{int} (resp. Σ_{inf}^{ext}, Σ_{sup}^{ext}) denote the inner (resp. outer) space

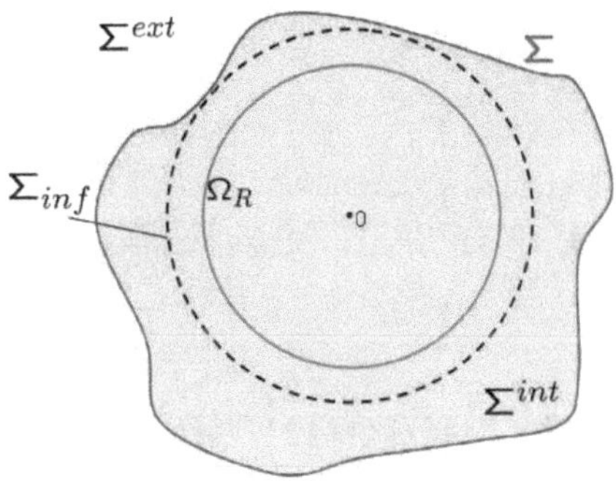

Figure 2.0.1: The geometrical concept. Ω_R denotes a sphere with radius R, Σ is a regular surface with Σ^{ext}, Σ^{int} denoting the exterior and interior space, respectively.

of the sphere Σ^{int} (resp. Σ^{ext}) around the origin with radius σ_{inf} (resp. σ_{sup}). Given a regular surface, there exists a positive constant R such that

$$R < \sigma_{\text{inf}} = \inf_{x \in \Sigma} |x| \leq \sup_{x \in \Sigma} |x| = \sigma_{\text{sup}}. \tag{2.0.33}$$

The set

$$\Sigma(\tau) = \{x \in \mathbb{R}^3 \,|\, x = y + \tau\nu(y),\, y \in \Sigma\}$$

generates a parallel surface which is exterior to Σ for $\tau > 0$ and interior for $\tau < 0$. From differential geometry (see [32]) it is well known that if $|\tau|$ is sufficiently small, then the regularity of Σ implies the regularity of $\Sigma(\tau)$ and the normal to one parallel surface is also a normal to the other. Moreover, it is easily seen (see [18]) that

$$\inf_{x,y \in \Sigma} |x + \tau\nu(x) - (y + \sigma\nu(y))| = |\tau - \sigma|$$

provided that $|\tau|$, $|\sigma|$ are sufficiently small.
In what follows every sphere $\Omega_R \subset \Sigma^{int}$ as indicated above will be called a Runge sphere.

Chapter 3

Spherical Polynomials

Our approach essentially follows [7, 8, 18, 21]. We start with the definition of homogeneous harmonic polynomials and introduce the addition theorem. Then we introduce scalar spherical harmonics as a restriction of the homogeneous harmonic polynomials to the sphere Ω. After that we present two types of vector spherical harmonics, the (Morse-Feshbach)-system $y_{n,m}^{(i)}$ and the (Edmonds-)system $u_{n,m}^{(i)}$, $n \in \mathbb{N}_{0_i}$; $m = 1, \ldots, 2n+1$; $i = 1, 2, 3$. Referring to Figure 1.0.1 this chapter deals with step (1) of the scalar and vector case.

3.1 Homogeneous and Homogeneous Harmonic Polynomials

First we follow the representation of homogeneous harmonic polynomials as introduced in [7, 14]. The set of all homogeneous polynomials of degree n (i.e., $H_n(\lambda x) = \lambda^n H_n(x)$, $\lambda \in \mathbb{R}, \lambda > 0$, and $x \in \mathbb{R}^3$) is denoted by Hom_n. If $H_n \in Hom_n$, then there exist real numbers $C_\alpha = C_{\alpha_1,\alpha_2,\alpha_3}$ such that

$$H_n(x) = H_n(x_1, x_2, x_3) = \sum_{[\alpha]=n} C_\alpha x^\alpha = \sum_{\alpha_1+\alpha_2+\alpha_3=n} C_{\alpha_1\alpha_2\alpha_3} x_1^{\alpha_1} x_2^{\alpha_2} x_3^{\alpha_3}. \quad (3.1.1)$$

Thus $Hom_n = \{H_n : \mathbb{R}^3 \to \mathbb{R} \mid H_n = \sum_{[\alpha]=n} C_\alpha x^\alpha\}$.

Remark 3.1.1. *The following properties are associated with the space Hom_n.*

(i) The set of monomials $x \mapsto x^\alpha$, $x \in \mathbb{R}^3$, $[\alpha] = n$ is a basis of Hom_n.

(ii) *The dimension of the space Hom_n is precisely the number of ways a triple can be chosen so that we have $[\alpha] = n$, i.e., the number of ways selecting 2 elements out of a collection of $n+2$. This means that the dimension $d(Hom_n)$ of Hom_n is equal to*

$$d(Hom_n) = \frac{(n+1)(n+2)}{2} = \binom{n+2}{2}.$$

Let $H_n(\nabla_x)$ be the differential operator associated to $H_n(x)$ (i.e., replace x^α formally by $(\nabla_x)^\alpha$ in the expression of $H_n(x)$) then

$$H_n(\nabla_x) = \sum_{\alpha_1+\alpha_2+\alpha_3=n} C_{\alpha_1\alpha_2\alpha_3} \frac{\partial^{[\alpha]}}{\partial x_1^{\alpha_1} \partial x_2^{\alpha_2} \partial x_3^{\alpha_3}} = \sum_{[\alpha]=n} C_\alpha (\nabla_x)^\alpha . \tag{3.1.2}$$

If such an operator is applied to a homogeneous polynomial U_n of the same degree

$$U_n(x) = \sum_{[\beta]=n} D_\beta \, x^\beta,$$

we obtain as result a real number:

$$\begin{aligned} &(H_n(\nabla_x))\, U_n(x) \\ &= \sum_{[\alpha]=n} \sum_{[\beta]=n} C_\alpha \, D_\beta \left(\frac{\partial}{\partial x_1}\right)^{\alpha_1} x_1^{\beta_1} \left(\frac{\partial}{\partial x_2}\right)^{\alpha_2} x_2^{\beta_2} \left(\frac{\partial}{\partial x_3}\right)^{\alpha_3} x_3^{\beta_3} \\ &= \sum_{[\alpha]=n} C_\alpha \, D_\alpha \, \alpha! \, . \end{aligned}$$

Clearly, we find

$$\begin{aligned} (H_n(\nabla_x))\, U_n(x) &= (U_n(\nabla_x))\, H_n(x), \\ (H_n(\nabla_x))\, H_n(x) &\geq 0. \end{aligned}$$

This enables us to introduce an inner product $(\cdot,\cdot)_{Hom_n}$ on the space Hom_n by letting

$$(H_n, U_n)_{Hom_n} = (H_n(\nabla_x))\, U_n(x). \tag{3.1.3}$$

The space Hom_n equipped with the inner product $(\cdot,\cdot)_{Hom_n}$ is a finite-dimensional Hilbert space. The set of monomials

$$\{x \mapsto (\alpha!)^{-1/2} x^\alpha \mid [\alpha] = n\}$$

forms an orthonormal system in the space Hom_n. For each $H_n \in Hom_n$ we have ([7, 14]) in connection with (3.1.2)

$$\begin{aligned} H_n(x) &= \sum_{[\alpha]=n} \frac{1}{\alpha!} (H_n(\nabla_y))\, y^\alpha\, x^\alpha \\ &= (H_n(\nabla_y))\, \frac{1}{n!} \sum_{[\alpha]=n} \frac{n!}{\alpha!}\, x^\alpha y^\alpha \\ &= (H_n(\nabla_y))\, \frac{(x \cdot y)^n}{n!} \\ &= \frac{1}{n!}\, (x \cdot \nabla_y)^n\, H_n(y). \end{aligned}$$

In other words,

$$H_n(x) = \left(\frac{(x \cdot\,)^n}{n!}, H_n \right)_{Hom_n}.$$

Theorem 3.1.1. *Hom_n equipped with the inner product $(\cdot,\cdot)_{Hom_n}$ is a finite-dimensional Hilbert space of dimension $\frac{(n+1)(n+2)}{2}$ with the reproducing kernel*

$$K_{Hom_n}(x,y) = \frac{(x \cdot y)^n}{n!}, \qquad x, y \in \mathbb{R}^3,$$

i.e.,

(i) for every fixed y, the function $K_{Hom_n}(\cdot, y)$ belongs to Hom_n,

(ii) for any $H_n \in Hom_n$ and any point x the reproducing property

$$H_n(x) = (K_{Hom_n}(x,\cdot), H_n)_{Hom_n}$$

is valid.

Let $\{H_{n,m}\}_{m=1,\ldots,d(Hom_n)}$, $\{U_{n,m}\}_{m=1,\ldots,d(Hom_n)}$ be two orthonormal systems in the space Hom_n:

$$\begin{aligned} (H_{n,m}, H_{n,k})_{Hom_n} &= \delta_{mk}, \\ (U_{n,m}, U_{n,k})_{Hom_n} &= \delta_{mk}, \end{aligned}$$

where δ_{mk} is the usual Kronecker symbol. Then, for $m = 1, ..., d(Hom_n)$, we have

$$\begin{aligned} H_{n,m} &= \sum_{k=1}^{d(Hom_n)} (H_{n,m}, U_{n,k})_{Hom_n} U_{n,k}, \\ U_{n,m} &= \sum_{k=1}^{d(Hom_n)} (U_{n,m}, H_{n,k})_{Hom_n} H_{n,k}. \end{aligned}$$

Therefore it follows that

$$\sum_{m=1}^{d(Hom_n)} H_{n,m}(x)\ H_{n,m}(y) = \sum_{m=1}^{d(Hom_n)} U_{n,m}(x) U_{n,m}(y).$$

Hence, in particular for the orthonormal system of monomials, we obtain the following result.

Theorem 3.1.2. *Let $\{H_{n,m}\}_{m=1,\dots,d(Hom_n)}$ be an orthonormal system in Hom_n. Then*

$$K_{Hom_n}(x,y) = \frac{(x\cdot y)^n}{n!} = \sum_{m=1}^{d(Hom_n)} H_{n,m}(x) H_{n,m}(y), \quad x, y \in \mathbb{R}^3.$$

$K_{Hom_n}(\cdot,\cdot)$ is the only reproducing kernel in Hom_n.

Suppose that there are given $d(Hom_n)$ points $x_1, ..., x_{d(Hom_n)} \in \mathbb{R}^3$ and $d(Hom_n)$-values $d_1, ...d_{d(Hom_n)} \in \mathbb{R}$. We are able to solve the Hom_n interpolation problem

$$\sum_{m=1}^{d(Hom_n)} b_m H_{n,m}(x_k) = d_k, \quad k = 1, ..., d(Hom_n),$$

if and only if the matrix

$$\mathbf{matr}_{\{x_1,\dots,x_{d(Hom_n)}\}}(H_{n,1}, \dots, H_{n,d(Hom_n)}) \tag{3.1.4}$$

$$= \begin{pmatrix} H_{n,1}(x_1) & \dots & H_{n,1}(x_{d(Hom_n)}) \\ \vdots & \ddots & \vdots \\ H_{n,d(Hom_n)}(x_1) & \dots & H_{n,d(Hom_n)}(x_{d(Hom_n)}) \end{pmatrix}$$

is non-singular. A system of $d(Hom_n)$ points $x_1, ..., x_{d(Hom_n)}$ is called a *fundamental system relative to Hom_n* if the matrix (3.1.4) is non-singular.

In what follows we guarantee the existence of a fundamental system relative to Hom_n (see for [31]).

Lemma 3.1.3. *There exists a system* $\{x_1, \ldots, x_{d(Hom_n)}\} \subset \mathbb{R}^3$ *such that (3.1.4) is non-singular.*

Proof. As orthonormal system, the functions $H_{n,1}, ..., H_{n,d(Hom_n)}$ are linearly independent. Hence, there exists a point x_1 for which

$$H_{n,1}(x_1) \neq 0.$$

Now, there must also be a point x_2 such that

$$\begin{vmatrix} H_{n,1}(x_1) & H_{n,1}(x_2) \\ H_{n,2}(x_1) & H_{n,2}(x_2) \end{vmatrix} \neq 0,$$

for else we would have a contradiction to the linear independence of $H_{n,1}, H_{n,2}$. In the same way the existence of a point x_3 can be deduced by the requirement

$$\begin{vmatrix} H_{n,1}(x_1) & H_{n,1}(x_2) & H_{n,1}(x_3) \\ H_{n,2}(x_1) & H_{n,2}(x_2) & H_{n,2}(x_3) \\ H_{n,3}(x_1) & H_{n,3}(x_2) & H_{n,3}(x_3) \end{vmatrix} \neq 0.$$

Finally, by induction, we obtain a system of points $x_1, ..., x_{d(Hom_n)}$ such that

$$\begin{vmatrix} H_{n,1}(x_1) & \ldots & H_{n,1}(x_{d(Hom_n)}) \\ \vdots & \ddots & \vdots \\ H_{n,d(Hom_n)}(x_1) & \ldots & H_{n,d(Hom_n)}(x_{d(Hom_n)}) \end{vmatrix} \neq 0,$$

i.e., $\{x_1, \ldots, x_{d(Hom_n)}\}$ constitutes a fundamental system relative to Hom_n. □

To every $H_n \in Hom_n$, there exist real numbers $b_1, ..., b_{d(Hom_n)}$ such that

$$H_n = \sum_{k=1}^{d(Hom_n)} b_k \, H_{n,k}.$$

Under the assumption that $\{x_1, ..., x_{d(Hom_n)}\}$ is a fundamental system relative to Hom_n, the linear equations

$$\sum_{j=1}^{d(Hom_n)} a_j \, H_{n,k}(x_j) = b_k, \quad k = 1, ..., d(Hom_n)\,, \tag{3.1.5}$$

are uniquely solvable in the unknowns $a_1, ..., a_{d(Hom_n)}$. Thus we obtain

$$H_n = \sum_{k=1}^{d(Hom_n)} \sum_{j=1}^{d(Hom_n)} a_j \, H_{n,k}(x_j) \, H_{n,k}.$$

Theorem 3.1.4. *Let $\{H_{n,m}\}_{m=1,\ldots,d(Hom_n)}$ be an orthonormal system in Hom_n. Assume that $\{x_k\}_{k=1,\ldots,d(Hom_n)}$ is a fundamental system relative to Hom_n. Then, each $H_n \in Hom_n$ is uniquely representable in the form*

$$H_n(x) = \sum_{m=1}^{d(Hom_n)} a_m \; K_{Hom_n}(x_m, x) = \sum_{m=1}^{d(Hom_n)} a_m \frac{(x_m \cdot x)^n}{n!}.$$

By $Harm_n$ we denote the space of all polynomials in Hom_n that are harmonic, i.e., fulfill Laplace's equation in three dimensions:

$$\mathrm{Harm}_n \;=\; \{H_n \in Hom_n \mid \Delta_x H_n(x) \;=\; 0,\; x \in \mathbb{R}^3\}\;.$$

For $n < 2$, of course, all homogeneous polynomials are harmonic.
Any homogeneous harmonic polynomial of degree n can be represented in the form

$$H_n(x) = H_n(x_1, x_2, x_3) = \sum_{j=0}^{n} x_3^j A_{n-j}(x_1, x_2), \tag{3.1.6}$$

where A_{n-j} is a homogeneous polynomial of degree $n-j$ in the variables x_1, x_2 given by the recursion relation

$$A_{n-j-2}(x_1, x_2) = -\frac{1}{(j+1)(j+2)}\left((\frac{\partial}{\partial x_1})^2 + (\frac{\partial}{\partial x_2})^2 \right) A_{n-j}(x_1, x_2),$$

for $j = 0, \ldots, n-2$. Therefore, all polynomials A_{n-j} are determined if we know A_n and A_{n-1}.

Theorem 3.1.5. *Let A_n and A_{n-1} be homogeneous polynomials of degree n and $n-1$ in $\mathbb{R}^2$, respectively. For $j = 0, ..., n-2$ we set recursively*

$$A_{n-j-2}(x_1, x_2) \;=\; -\frac{1}{(j+1)(j+2)}\left(\left(\frac{\partial}{\partial x_1}\right)^2 + \left(\frac{\partial}{\partial x_2}\right)^2 \right) A_{n-j}(x_1, x_2). \tag{3.1.7}$$

Then $H_n : \mathbb{R}^3 \to \mathbb{R}$ given by

$$H_n(x_1, x_2, x_3) \;=\; \sum_{j=0}^{n} x_3^j A_{n-j}(x_1, x_2)$$

is a homogeneous harmonic polynomial of degree n in $\mathbb{R}^3$, i.e. $H_n \in Harm_n$. The number of linearly independent homogeneous harmonic polynomials is equal to the number of coefficients of A_n and A_{n-1}, i.e.

$$d(Harm_n) = n + n + 1 = 2n + 1.$$

Assume that $n \geq 2$. Let H_{n-2} be a homogeneous polynomial of degree $n-2$, i.e. $H_{n-2} \in \mathrm{Hom}_{n-2}$. Then, for each homogeneous harmonic polynomial K_n, we have

$$(|\cdot|^2 H_{n-2}, K_n)_{Hom_n} = (H_{n-2}(\nabla_x))\Delta_x K_n(x) = 0 \ .$$

This means $|\cdot|^2 H_{n-2}$ is orthogonal to K_n in the sense of the inner product $(\cdot,\cdot)_{Hom_n}$. Conversely, suppose that $K_n \in Hom_n$ is orthogonal to all elements L_n of the form

$$L_n(x) \ = \ |x|^2 H_{n-2}(x) \ , \quad H_{n-2} \in \mathrm{Hom}_{n-2}.$$

Then it follows that

$$0 = (|\cdot|^2 H_{n-2}, K_n)_{Hom_n} = (H_{n-2}(\nabla_x))\Delta_x K_n(x) = (H_{n-2}, \Delta K_n)_{\mathrm{Hom}_{n-2}}$$

for all $H_{n-2} \in \mathrm{Hom}_{n-2}$. This is true only if $\Delta K_n = 0$, i.e. K_n is a homogeneous harmonic polynomial.

Theorem 3.1.6. *(Decomposition theorem of Hom_n) Hom_n, $n \geq 2$, is the orthogonal direct sum of Harm_n and $\mathrm{Harm}_n^{\perp}$, where $\mathrm{Harm}_n^{\perp} = |\cdot|^2 Hom_{n-2}$ is the space of all L_n with $L_n(x) = |x|^2 H_{n-2}(x)$, $H_{n-2} \in Hom_{n-2}$. Consequently, each homogeneous polynomial H_n of degree n can be uniquely decomposed in the form*

$$H_n(x) \ = \ K_n(x) + |x|^2 H_{n-2}(x) \ ,$$

where K_n is a homogeneous harmonic polynomial of degree n and H_{n-2} is a homogeneous polynomial of degree $n-2$.

Denote by $Proj_{\mathrm{Harm}_n}$ and $Proj_{\mathrm{Harm}_n^{\perp}}$ the projection operators in Hom_n onto Harm_n and $\mathrm{Harm}_n^{\perp}$, respectively. Then

$$H_n \ = \ Proj_{\mathrm{Harm}_n} H_n + Proj_{\mathrm{Harm}_n^{\perp}} H_n \ .$$

In other words,

$$\begin{aligned} K_n(x) \ &= \ Proj_{\mathrm{Harm}_n} H_n(x), \\ |x|^2 H_{n-2}(x) \ &= \ Proj_{\mathrm{Harm}_n^{\perp}} H_n(x) \ . \end{aligned}$$

For all $H_n, U_n \in Hom_n$,

$$(Proj_{\mathrm{Harm}_n} H_n, U_n)_{Hom_n} \ = \ (H_n, Proj_{\mathrm{Harm}_n} U_n)_{Hom_n} \ .$$

Moreover, we have $Proj_{\mathrm{Harm}_n} H_n = \mathrm{Proj}_{\mathrm{Harm}_n} K_n = K_n$. Observe that

$$\begin{aligned} d(\mathrm{Harm}_n) &= d(Hom_n) - d(\mathrm{Harm}_n^{\perp}) \\ &= d(Hom_n) - d(\mathrm{Hom}_{n-2}) \\ &= \binom{n+2}{2} - \binom{n}{2} = 2n+1. \end{aligned}$$

If we apply Theorem 3.1.6 recursively to H_{n-2}, H_{n-4}, ..., we obtain the following result.

Theorem 3.1.7. *Each homogeneous polynomial of degree n can be uniquely decomposed in the form*

$$H_n(x) = \sum_{i=0}^{[n/2]} |x|^{2i} K_{n-2i}(x), \quad K_{n-2i} \in Harm_{n-2i}, \quad x \in \mathbb{R}^3, \tag{3.1.8}$$

where

$$\left[\frac{n}{2}\right] = \frac{1}{2}\left(n - \frac{1}{2}(1-(-1)^n)\right).$$

In other words, Hom_n admits the direct sum decomposition

$$Hom_n(\mathbb{R}^3) = \bigoplus_{i=0}^{[n/2]} |\cdot|^{2i} Harm_{n-2i}(\mathbb{R}^3).$$

This result gives rise to the following corollary.

Corollary 3.1.8. *For $n \in \mathbb{N}_0$*

$$Hom_n(\mathbb{R}^3)|_\Omega = Hom_n(\Omega) = \bigoplus_{i=0}^{[n/2]} Harm_{n-2i}(\mathbb{R}^3)|_\Omega.$$

Since the space $\mathrm{P}ol_{0,\ldots,n}(\mathbb{R}^3)$ of polynomials in three variables of degree $\leq n$ can be written as direct sum decomposition of $Hom_n(\mathbb{R}^3)$ and $Hom_{n-1}(\mathbb{R}^3)$, when restricted to Ω, i.e.,

$$\mathrm{P}ol_{0,\ldots,n}(\mathbb{R}^3)|_\Omega = (Hom_n(\mathbb{R}^3)|_\Omega) \oplus (\mathrm{H}om_{n-1}(\mathbb{R}^3)|_\Omega).$$

We finally obtain the following.

Corollary 3.1.9. *For $n \in \mathbb{N}_0$*

$$Pol_{0,\ldots,n}(\mathbb{R}^3)|_\Omega = \bigoplus_{i=0}^{n} Harm_i(\mathbb{R}^3)|_\Omega.$$

In other words, the restriction to the unit sphere Ω of any polynomial of three variables is a sum of restrictions to Ω of homogeneous harmonic polynomials.

Addition Theorem for Homogeneous Harmonic Polynomials

We are now interested in giving the explicit representation of the orthogonal projection $Proj_{\mathrm{Harm}_n} H_n$ of a given homogeneous polynomial H_n. For that purpose we need some preliminaries. By induction we are able to prove that for $i = 1, 2, 3$ and $|x| \neq 0$ (see for [26])

$$\begin{aligned} &\left(\frac{\partial}{\partial x_i}\right)^n \frac{1}{|x|} \\ &= (-1)^n \frac{(2n)!}{n! 2^n} \frac{1}{|x|^{2n+1}} \left(\sum_{s=0}^{[n/2]} (-1)^s \frac{n!(2n-2s)!}{(2n)!(n-s)!s!} |x|^{2s} \Delta^s \right) x_i^n. \end{aligned}$$

In other words, we find

$$\begin{aligned} &(\varepsilon^i \cdot \nabla_x)^n \frac{1}{|x|} \\ &= (-1)^n \frac{(2n)!}{n! 2^n} \frac{1}{|x|^{2n+1}} \left(\sum_{s=0}^{[n/2]} (-1)^s \frac{n!(2n-2s)!}{(2n)!(n-s)!s!} |x|^{2s} \Delta^s \right) (\varepsilon^i \cdot x)^n, \end{aligned}$$

$i = 1, 2, 3$. Since the differential operator Δ is invariant with respect to orthogonal transformations it is easy to see that

$$\begin{aligned} &(y \cdot \nabla_x)^n \frac{1}{|x|} \\ &= (-1)^n \frac{(2n)!}{n! 2^n} \frac{1}{|x|^{2n+1}} \left(\sum_{s=0}^{[n/2]} (-1)^s \frac{n!(2n-2s)!}{(2n)!(n-s)!s!} |x|^{2s} \Delta^s \right) (y \cdot x)^n \end{aligned}$$

is valid for every $y \in \mathbb{R}^3$. Now, as we have seen in Theorem 3.1.4, each $H_n \in Hom_n$ may be represented in the form

$$H_n(x) = \sum_{m=1}^{d(Hom_n)} c_m (x_m \cdot x)^n, \quad x \in \mathbb{R}^3,$$

where c_m, $m = 1, \ldots, d(Hom_n)$, are suitable coefficients and $x_1, \ldots, x_{d(Hom_n)}$ is a fundamental system relative to Hom_n.

Consequently, we have the following result.

Theorem 3.1.10. *Let H_n be a homogeneous polynomial of degree n. Then, for each $x \in \mathbb{R}^3$, $|x| \neq 0$,*

$$\begin{aligned}&(H_n(\nabla_x))\frac{1}{|x|}\\ &= (-1)^n \frac{(2n)!}{n!2^n}\frac{1}{|x|^{2n+1}}\left(\sum_{s=0}^{[n/2]}(-1)^s\frac{n!(2n-2s)!}{(2n)!(n-s)!s!}|x|^{2s}\Delta^s\right)H_n(x).\end{aligned}$$

From the considerations given above it follows that

$$(H_n(\nabla_x))\frac{1}{|x|} = (K_n(\nabla_x))\frac{1}{|x|} + (H_{n-2}(\nabla_x))\Delta_x\frac{1}{|x|}, \quad |x| \neq 0.$$

Thus, in connection with

$$\begin{aligned}\Delta_x\frac{1}{|x|} &= 0, \quad |x| \neq 0,\\ \Delta_x K_n(x) &= 0, \quad x \in \mathbb{R}^3,\end{aligned}$$

we obtain for $|x| \neq 0$

$$(H_n(\nabla_x))\frac{1}{|x|} = (K_n(\nabla_x))\frac{1}{|x|} = (-1)^n\frac{(2n)!}{n!2^n}\frac{1}{|x|^{2n+1}}K_n(x). \tag{3.1.9}$$

Therefore, by comparison of (3.1.9) and Theorem 3.1.10, we get the following lemma.

Lemma 3.1.11. *Let H_n be a homogeneous polynomial of degree n. Then*

$$Proj_{Harm_n} H_n(x) = \left(\sum_{s=0}^{[n/2]}(-1)^s\frac{n!(2n-2s)!}{(2n)!(n-s)!s!}|x|^{2s}\Delta^s\right)H_n(x).$$

Observing

$$\Delta_x(x\cdot y)^n = n(n-1)|y|^2(x\cdot y)^{n-2}, \quad y \in \mathbb{R}^3,$$

we obtain, in particular,

$$
\begin{aligned}
&Proj_{\mathrm{Harm}_n}(\frac{(x\cdot y)^n}{n!})\\
&= \frac{1}{n!}\sum_{s=0}^{[n/2]}(-1)^s\frac{(2n-2s)!(n!)^2}{(n-2s)!(n-s)!s!(2n)!}|x|^{2s}|y|^{2s}(x\cdot y)^{n-2s}.
\end{aligned}
$$

Thus, we find by using $x = |x|\xi$, $y = |y|\eta$, $(\xi,\eta) \in \Omega^2$, the equation

$$
\begin{aligned}
&Proj_{\mathrm{Harm}_n}\left(\frac{(x\cdot y)^n}{n!}\right)\\
&= \frac{(2n+1)2^n\cdot n!}{(2n+1)!}(|x|\;|y|)^n\sum_{s=0}^{[n/2]}(-1)^s\frac{(2n-2s)!}{2^n(n-2s)!(n-s)!s!}(\xi\cdot\eta)^{n-2s}.
\end{aligned}
$$

Suppose that $\{H_{n,m}\}_{m=1,\ldots,d(\mathrm{Harm}_n)}$ is an orthonormal system in Harm_n with respect to $(\cdot,\cdot)_{Hom_n}$. Let $\{U_{n,m}\}_{m=1,\ldots,d(Hom_n)-d(\mathrm{Harm}_n)}$ be an orthonormal system in $\mathrm{Harm}_n^{\perp}$. Then the union of both systems

$$
\{H_{n,m}\}_{m=1,\ldots,d(\mathrm{Harm}_n)} \cup \{U_{n,m}\}_{m=1,\ldots,d(Hom_n)-d(\mathrm{Harm}_n)}
$$

forms an orthonormal system in Hom_n. Therefore it follows that

$$
\begin{aligned}
&\frac{(x\cdot y)^n}{n!} \qquad (3.1.10)\\
&= \sum_{m=1}^{d(\mathrm{Harm}_n)} H_{n,m}(x)\;H_{n,m}(y) + \sum_{m=1}^{d(Hom_n)-d(\mathrm{Harm}_n)} U_{n,m}(x)\;U_{n,m}(y)
\end{aligned}
$$

for any pair $x, y \in \mathbb{R}^3$. On the one hand, in view of the definition of the projection operator $Proj_{\mathrm{Harm}_n}$, we get

$$
\begin{aligned}
&Proj_{\mathrm{Harm}_n}\left(\sum_{m=1}^{d(\mathrm{Harm}_n)} H_{n,m}(x)H_{n,m}(y) + \sum_{m=1}^{d(Hom_n)-d(\mathrm{Harm}_n)} U_{n,m}(x)U_{n,m}(y)\right)\\
&= \sum_{m=1}^{d(\mathrm{Harm}_n)} H_{n,m}(x)H_{n,m}(y). \qquad (3.1.11)
\end{aligned}
$$

On the other hand, as we have shown above,

$$
\begin{aligned}
&Proj_{\mathrm{Harm}_n}(\frac{(x\cdot y)^n}{n!}) \qquad (3.1.12)\\
&= \frac{(2n+1)2^n n!}{(2n+1)!}|x|^n|y|^n\sum_{s=0}^{[n/2]}(-1)^s\frac{(2n-2s)!}{2^n(n-2s)!(n-s)!s!}(\xi\cdot\eta)^{n-2s}.
\end{aligned}
$$

By comparison of (3.1.11) and (3.1.12) we obtain the *addition theorem of homogeneous harmonic polynomials in* $\mathbb{R}^3$.

Theorem 3.1.12. *Let* $\{H_{n,m}\}_{m=1,\dots,d(Harm_n)}$, $d(Harm_n) = 2n+1$, *be an orthonormal system in* $Harm_n$ *with respect to* $(\cdot,\cdot)_{Hom_n}$. *Then, for* $x, y \in \mathbb{R}^3$, $x = |x|\xi$, $y = |y|\eta$, *we have*

$$\sum_{j=1}^{2n+1} H_{n,m}(x)\ H_{n,m}(y) = \frac{2^n n!}{(2n)!}|x|^n|y|^n P_n(\xi\cdot\eta),$$

where we have used the abbreviation

$$P_n(t) \;=\; \sum_{s=0}^{[n/2]}(-1)^s \frac{(2n-2s)!}{2^n(n-2s)!(n-s)!s!}t^{n-2s}, \quad t\in[-1,1].$$

Next we discuss the important question of how, for any pair of elements $H_n \in Harm_n$, $K_n \in \mathrm{Harm}_n$, the inner product $(\cdot,\cdot)_{Hom_n}$ defined by (3.1.3) is related to the (usually used) inner product $(\cdot,\cdot)_{\mathcal{L}^2(\Omega)}$.

Theorem 3.1.13. *For* $H_m \in Harm_m, K_n \in Harm_n$,

$$(H_m, K_n)_{\mathcal{L}^2(\Omega)} \;=\; \frac{\delta_{nm}}{\mu_n}(H_m(\nabla_x))K_n(x),$$

where μ_n *is given by*

$$\mu_n = \frac{(2n+1)!}{4\pi 2^n n!} = \frac{1\cdot 3\cdot\ldots\cdot(2n+1)}{4\pi} = \frac{(2n+1)!!}{4\pi}. \tag{3.1.13}$$

Proof. By virtue of the fundamental theorem of potential theory (see, for example, [27])

$$K_n(x) \;=\; \frac{1}{4\pi}\int_\Omega \left\{\frac{1}{|x-y|}\frac{\partial}{\partial\nu_y}K_n(y) - K_n(y)\frac{\partial}{\partial\nu_y}\frac{1}{|x-y|}\right\}d\omega(y)$$

for all $x\in\mathbb{R}^3$ with $|x|<1$, where $\partial/\partial\nu$ denotes the derivative in the direction of the outer normal to Ω. Therefore we find

$$\begin{aligned}(H_m(\nabla_x))K_n(x) = \frac{1}{4\pi}\int_\Omega &\left\{(H_m(\nabla_x))\frac{1}{|x-y|}\frac{\partial}{\partial\nu_y}K_n(y)\right.\\ &- \left. K_n(y)\frac{\partial}{\partial\nu_y}(H_m(\nabla_x))\frac{1}{|x-y|}\right\}d\omega(y).\end{aligned} \tag{3.1.14}$$

For $x\neq y$ we get from (3.1.9)

$$(H_m(\nabla_x))\frac{1}{|x-y|} \;=\; (-1)^m\frac{(2m)!}{m!2^m}\frac{H_m(x-y)}{|x-y|^{2m+1}}.$$

Because H_m is homogeneous, this is equivalent to

$$(H_m(\nabla_x))\frac{1}{|x-y|} = \frac{(2m)!}{m!2^m}\frac{H_m(y-x)}{|x-y|^{2m+1}}. \tag{3.1.15}$$

Inserting (3.1.15) into (3.1.14) gives

$$\begin{aligned}(H_m(\nabla_x))K_n(x) &= \frac{(2m)!}{(m!)2^m}\frac{1}{4\pi}\int_\Omega\left\{\frac{H_m(y-x)}{|x-y|^{2m+1}}\frac{\partial}{\partial\nu_y}K_n(y)\right.\\ &\quad \left.-K_n(y)\frac{\partial}{\partial\nu_y}\frac{H_m(y-x)}{|x-y|^{2m+1}}\right\} d\omega(y).\end{aligned}$$

It is easy to see that for $m \neq n$

$$(H_m(\nabla_x))K_n(x)\mid_{x=0} = 0,$$

while for $m = n$

$$(H_m(\nabla_x))K_n(x)\mid_{x=0} = (H_m(\nabla_x))K_n(x) = (H_m, K_n)_{Hom_n}.$$

Therefore we obtain

$$\frac{1}{4\pi}\int_\Omega\left\{\frac{H_m(y)}{|y|^{2m+1}}\frac{\partial}{\partial\nu_y}K_n(y) - K_n(y)\frac{\partial}{\partial\nu_y}\frac{H_m(y)}{|y|^{2m+1}}\right\} d\omega(y) \tag{3.1.16}$$

$$= \begin{cases} 0 & \text{for} \quad m\neq n \\ \left(\frac{2^m m!}{(2m)!}\right)(H_m, K_n)_{Hom_n} & \text{for} \quad m = n \end{cases}.$$

Since the normal derivatives of K_n and H_m are equal to

$$\frac{\partial}{\partial r}K_n(r\xi)\mid_{r=1} = nK_n(\xi)\ , \qquad \frac{\partial}{\partial r}H_m(r\xi)\mid_{r=1} = mH_m(\xi), \tag{3.1.17}$$

respectively, it follows that

$$\begin{aligned}&\frac{1}{4\pi}\int_\Omega\left\{\frac{H_m(y)}{|y|^{2m+1}}\frac{\partial}{\partial\nu_y}K_n(y) - K_n(y)\frac{\partial}{\partial\nu_y}\frac{H_m(y)}{|y|^{2m+1}}\right\} d\omega(y)\\ &= \frac{1}{4\pi}\int_\Omega \{nH_m(\xi)K_n(\xi) + (m+1)H_m(\xi)K_n(\xi)\}\, d\omega(\xi)\\ &= \frac{n+m+1}{4\pi}\int_\Omega H_m(\xi)K_n(\xi)\, d\omega(\xi). \end{aligned} \tag{3.1.18}$$

Thus, by combination of (3.1.16) and (3.1.18), we finally obtain the desired result stated in Theorem 3.1.13. □

Thus, to any orthonormal system $\{H_{n,m}\}_{m=1,\dots,2n+1}$ in $Harm_n$ with respect to $(\cdot,\cdot)_{Hom_n}$ there corresponds the $\mathcal{L}^2(\Omega)$-orthonormal system $\{\sqrt{\mu_n}H_{n,m}\}_{m=1,\dots,2n+1}$, and vice versa.

Finally, we are led to the following reformulation of the addition theorem.

Theorem 3.1.14. *Let $\{H_{n,m}\}_{m=1,\dots,2n+1}$ be an orthonormal system in $Harm_n$ with respect to $(\cdot,\cdot)_{Hom_n}$. Then, for $x, y \in \mathbb{R}^3$, we have*

$$\sum_{m=1}^{2n+1} \sqrt{\mu_n}H_{n,m}(x)\ \sqrt{\mu_n}H_{n,m}(y) = \frac{2n+1}{4\pi}|x|^n|y|^n P_n(\xi\cdot\eta),$$

where $x = |x|\xi,\ y = |y|\eta;\ (\xi,\eta) \in \Omega^2$ and μ_n is defined by (3.1.13)

3.2 Scalar Spherical Harmonics

Spherical harmonics are the analogues of trigonometric functions on spheres (see, e.g., [7, 14, 31]). One possibility to define them is as restrictions of homogeneous harmonic polynomials to the unit sphere. Other possibilities are, e.g., given by the use of the Beltrami operator or with the help of Legendre polynomials (see Definition 2.0.1). Every spherical harmonic, which is invariant under orthogonal transformations can be represented by a Legendre polynomial. This relationship is reflected by the addition theorem for scalar spherical harmonics, which will also be given in this section.
We provide first the definition of spherical harmonics and the development of their properties in $\mathbb{R}^3$.

Definition 3.2.1. *Let $H_n : \mathbb{R}^3 \to \mathbb{R}$ be a homogeneous harmonic polynomial in three variables of degree n with $n \in \mathbb{N}_0$. We call its restriction to the unit sphere*

$$Y_n = H_n|_\Omega$$

a spherical harmonic of degree n. The space of all spherical harmonics of degree n is denoted by $Harm_n(\Omega)$. Additionally the space of all spherical harmonics of degree less or equal to n will be denoted $Harm_{0,\dots,n}$.

We now state some useful and important facts about spherical harmonics (see for more details, e.g., [7, 14, 31]):

1. Spherical harmonics of different degrees are orthogonal with respect to the $\mathcal{L}^2$-scalar product, i.e., for every $Y_n \in Harm_n$ and every $Y_m \in Harm_m$ we have
$$(Y_n, Y_m)_{\mathcal{L}^2(\Omega)} = \int_\Omega Y_n(\eta)Y_m(\eta)d\omega(\eta) = 0, \quad n \neq m, \quad \eta \in \Omega.$$
2. $Harm_n(\Omega)$ is of dimension $dim(Harm_n) = 2n+1$.
3. From the last property we can deduce that
$$Harm_{0,\dots,n}(\Omega) = \bigoplus_{m=0}^{n} Harm_m(\Omega), \tag{3.2.1}$$
hence,
$$dim(Harm_{0,\dots,n}(\Omega)) = \sum_{m=0}^{n}(2m+1) = (n+1)^2.$$

Applying the Laplace operator to $H_n \in Harm_n(\mathbb{R}^3)$ and observing that $H_n(x) = r^n Y_n(\xi)$ with $Y_n \in Harm_n(\Omega)$ and $x = r\xi \in \mathbb{R}^3$, $r = |x|$, $\xi \in \Omega$, we get another possibility to introduce spherical harmonics, namely as eigenfunctions of the Beltrami operator.

Lemma 3.2.1. *Any spherical harmonic Y_n, $n \in \mathbb{N}_0$, is an infinitely often differentiable eigenfunction of the Beltrami operator corresponding to the eigenvalue $-n(n+1)$. More explicitly,*
$$(\Delta^*_\xi - (\Delta^*)^\wedge(n))Y_n(\xi) = 0, \quad \xi \in \Omega, \quad Y_n \in Harm_n,$$
where the 'spherical symbol' $\{(\Delta^*)^\wedge(n)\}_{n\in\mathbb{N}_0}$ of the Beltrami operator Δ^* is given by $(\Delta^*)^\wedge(n) = -n(n+1)$, $n \in \mathbb{N}_0$.

Remark 3.2.1. *With $Y_{n,m}$ we denote a member of an orthonormal system $\{Y_{n,1}, \dots, Y_{n,2n+1}\}$ in $Harm_n$ with respect to $(\cdot,\cdot)_{\mathcal{L}^2(\Omega)}$.*

In the following we list the completeness and closure properties for spherical harmonics in $\mathcal{C}(\Omega)$ and $\mathcal{L}^2(\Omega)$. They enable us to expand scalar functions $F \in \mathcal{L}^2(\Omega)$ (observe that $\mathcal{C}(\Omega) \subset \mathcal{L}^2(\Omega)$) into a Fourier series of spherical harmonics.

The system $\{Y_{n,m}\}_{\substack{n\in\mathbb{N}_0;\\ m=1,\dots,2n+1}}$ is closed in $(\mathcal{C}(\Omega), \|\cdot\|_{\mathcal{C}(\Omega)})$. That means that for each $F \in \mathcal{C}(\Omega)$ and any $\varepsilon > 0$ there exists a linear combination
$$\sum_{n=0}^{N_\varepsilon}\sum_{m=1}^{2n+1} d_{n,m}Y_{n,m}, \tag{3.2.2}$$

such that

$$\left\| F - \sum_{n=0}^{N_\varepsilon} \sum_{m=1}^{2n+1} d_{n,m} Y_{n,m} \right\|_{\mathcal{C}(\Omega)} \leq \varepsilon. \tag{3.2.3}$$

Further, the system $\{Y_{n,m}\}_{\substack{n\in\mathbb{N}_0;\\ m=1,\dots,2n+1}}$ is closed in the space $\mathcal{C}(\Omega)$ with respect to $\|\cdot\|_{\mathcal{L}^2(\Omega)}$, that is for any given $\varepsilon > 0$ and any given $F \in \mathcal{C}(\Omega)$, there exists a linear combination

$$\sum_{n=0}^{N_\varepsilon} \sum_{m=1}^{2n+1} d_{n,m} Y_{n,m},$$

such that

$$\left\| F - \sum_{n=0}^{N_\varepsilon} \sum_{m=1}^{2n+1} d_{n,m} Y_{n,m} \right\|_{\mathcal{L}^2(\Omega)} \leq \varepsilon. \tag{3.2.4}$$

From (2.0.4) we know that $\mathcal{C}(\Omega)$ is dense in $\mathcal{L}^2(\Omega)$, so we deduce that for each $F \in \mathcal{L}^2(\Omega)$ there exists a $G \in \mathcal{C}(\Omega)$ lying arbitrarily close to F in the $\mathcal{L}^2$–topology. In connection with 3.2.4 we thus can formulate:

Corollary 3.2.2. *The system* $\{Y_{n,m}\}_{\substack{n\in\mathbb{N}_0;\\ m=1,\dots,2n+1}}$ *is closed in the space* $\mathcal{L}^2(\Omega)$ *with respect to* $\|\cdot\|_{\mathcal{L}^2(\Omega)}$.

Lemma 3.2.3. *Let* $F \in \mathcal{L}^2(\Omega)$ *and let* $\{Y_{n,m}\}_{\substack{n\in\mathbb{N}_0;\\ m=1,\dots,2n+1}}$ *be an* $\mathcal{L}^2(\Omega)$*- orthonormal system, then*

$$\left\| F - \sum_{n=0}^{a} \sum_{m=1}^{2n+1} F^{\wedge}(n,m) Y_{n,m} \right\|_{\mathcal{L}^2(\Omega)} = \inf_{Y \in Harm_{0,\dots,a}} \|F - Y\|_{\mathcal{L}^2(\Omega)},$$

where

$$F^{\wedge}(n,m) = \int_\Omega F(\eta) Y_{n,m}(\eta) d\omega(\eta).$$

Thus, any element F of class $\mathcal{L}^2(\Omega)$ allows a representation by its Fourier series

$$F = \sum_{n=0}^{\infty} \sum_{m=1}^{2n+1} F^{\wedge}(n,m) Y_{n,m},$$

where

$$F^{\wedge}(n,m) = \int_\Omega F(\eta) Y_{n,m}(\eta) d\omega(\eta)$$

are the so-called Fourier- (or orthogonal) coefficients of F.

Most of the results of this section are summarized in the fundamental theorem of spherical harmonic expansion.

Theorem 3.2.4. *The following seven statements are equivalent.*

1. $\{Y_{n,m}\}_{\substack{n\in\mathbb{N}_0;\\ m=1,\ldots,2n+1}}$ *is closed in* $\mathcal{L}^2(\Omega)$ *(closure property).*
2. *The orthogonal expansion of any* $F \in \mathcal{L}^2(\Omega)$ *converges in norm to* F, *i.e.,*
$$\lim_{a\to\infty}\left\| F - \sum_{n=0}^{a}\sum_{m=1}^{2n+1}(F,Y_{n,m})_{\mathcal{L}^2(\Omega)}Y_{n,m}\right\|_{\mathcal{L}^2(\Omega)} = 0.$$
3. *Parseval's identity holds true, i.e., for any* $F \in \mathcal{L}^2(\Omega)$ *we have*
$$\|F\|^2_{\mathcal{L}^2(\Omega)} = (F,F)_{\mathcal{L}^2(\Omega)} = \sum_{n=0}^{\infty}\sum_{m=1}^{2n+1}|(F,Y_{n,m})_{\mathcal{L}^2(\Omega)}|^2.$$
4. *Extended Parseval's identity holds true, i.e., for any* $F, G \in \mathcal{L}^2(\Omega)$ *we have*
$$(F,G)_{\mathcal{L}^2(\Omega)} = \sum_{n=0}^{\infty}\sum_{m=1}^{2n+1}(F,Y_{n,m})_{\mathcal{L}^2}(G,Y_{n,m})_{\mathcal{L}^2}.$$
5. *There is no strictly larger orthonormal system containing the orthonormal system* $\{Y_{n,m}\}_{\substack{n\in\mathbb{N}_0;\\ m=1,\ldots,2n+1}}$.
6. *The system* $\{Y_{n,m}\}_{\substack{n\in\mathbb{N}_0;\\ m=1,\ldots,2n+1}}$ *fulfills the completeness property. That is,* $F \in \mathcal{L}^2(\Omega)$ *and* $(F,Y_{n,m})_{\mathcal{L}^2(\Omega)} = 0$ *for all* $n \in \mathbb{N}_0$ *and* $m = 1,\ldots,2n+1$, *implies* $F = 0$.
7. *Any element* $F \in \mathcal{L}^2(\Omega)$ *is uniquely determined by its orthogonal coefficients. That means if* $(F,Y_{n,m})_{\mathcal{L}^2(\Omega)} = (G,Y_{n,m})_{\mathcal{L}^2(\Omega)}$ *for all* $n \in \mathbb{N}_0$ *and* $m = 1,\ldots,2n+1$, *implies* $F = G$.

The addition theorem, introduced next, builds the bridge between zonal, i.e., radial basis functions and spherical harmonics.

Theorem 3.2.5. *Let $\{Y_{n,m}\}$, $n \in \mathbb{N}_0$; $m = 1, \ldots, 2n+1$, be a system of orthonormal spherical harmonics of degree n, and let P_n be the Legendre polynomial of degree n. Then, for all ξ, $\eta \in \Omega$*

$$\sum_{m=1}^{2n+1} Y_{n,m}(\xi) Y_{n,m}(\eta) = \frac{2n+1}{4\pi} P_n(\xi \cdot \eta). \tag{3.2.5}$$

This immediately leads to

$$\sum_{m=1}^{2n+1} Y_{n,m}(\xi) Y_{n,m}(\xi) = \sum_{m=1}^{2n+1} (Y_{n,m}(\xi))^2 = \frac{2n+1}{4\pi}. \tag{3.2.6}$$

Another important theorem that sets radial basis functions and spherical harmonics into relation is the following formula of Funk-Hecke.

Theorem 3.2.6. *Let $G \in \mathcal{L}^1[-1,1]$ and let P_n be the Legendre polynomial. Then for all $(\xi, \eta) \in \Omega^2$ and $n \in \mathbb{N}_0$,*

$$\int_\Omega G(\xi \cdot \zeta) P_n(\eta \cdot \zeta) d\omega(\zeta) = G^\wedge(n) P_n(\xi \cdot \eta),$$

where $G^\wedge(n)$ is the Legendre coefficient of G, i.e.,

$$G^\wedge(n) = (G, P_n)_{\mathcal{L}^2[-1,1]}.$$

Moreover, if Y_n is a spherical harmonic of degree n, then

$$\int_\Omega G(\xi \cdot \eta) Y_n(\eta) d\omega(\eta) = G^\wedge(n) Y_n(\xi).$$

That means that the spherical harmonics Y_n are the eigenfunctions of the above integral operator corresponding to the eigenvalue $G^\wedge(n)$. Thus, the last theorem establishes a connection between spherical harmonics and radial basis functions and founds the basis for the introduction of spherical singular integrals and spherical wavelets (see [14], [17] and [22], for example).

3.3 Vector Spherical Harmonics

We will give now an overview on two types of vector spherical harmonics: the (Morse-Feshbach-)vector spherical harmonics $y^{(i)}_{n,m}$ and the (Edmonds-)vector spherical harmonics $u^{(i)}_{n,m}$, $n \in \mathbb{N}_{0_i}$; $m = 1, \ldots, 2n+1$; $i = 1, 2, 3$. Both types are derived from the scalar spherical harmonics by using different classes of differential operators which are applied on $Y_n \in Harm_n(\Omega)$. More information can be found in [14, 30].

(Morse-Feshbach-)Vector Spherical Harmonics $y^{(i)}_{n,m}$

In order to construct vector spherical harmonics $y^{(i)}_{n,m}$, $n \in \mathbb{N}_{0_i}$; $m = 1, \ldots, 2n+1$; $i = 1, 2, 3$, which on the one hand separate into normal and tangential parts and on the other hand lead to an orthonormal basis of $l^2(\Omega)$ we use the operators $o^{(i)}$, $i = 1, 2, 3$, as given by (2.0.27).

Definition 3.3.1. *Let Y_n be of class $Harm_n(\Omega)$, $i \in 1, 2, 3$ and $n \in \mathbb{N}_{0_i}$. Then we call the vector field*

$$y^{(i)}_n(\xi) = o^{(i)}_\xi Y_n(\xi), \quad \xi \in \Omega,$$

(Morse-Feshbach-)vector spherical harmonic of degree n and type i.

Further, we denote the space of all vector spherical harmonics of degree n and type i by $harm^{(i)}_n(\Omega)$ and let

$$\begin{aligned} harm_0 =& harm^{(1)}_0 \\ harm_n(\Omega) =& \bigoplus_{i=1}^{3} harm^{(i)}_n(\Omega), \quad n \geq 1. \end{aligned}$$

Regarding Remark 2.0.1 and the definitions for the operators $O^{(i)}$, $i = 1, 2, 3$, given by (2.0.30), we can orthogonally split the space $c^{(\infty)}(\Omega)$ as follllows:

$$c^{(\infty)}(\Omega) = \bigoplus_{i=1}^{3} c^{(\infty)}_{(i)}(\Omega),$$

where

$$c^{(\infty)}_{(i)}(\Omega) = \{f \in c^{(\infty)}(\Omega) \,|\, O^{(k)} f = 0, \quad i \neq k\}.$$

Then,

$$\begin{aligned} c_{nor}^{(\infty)}(\Omega) &= c_{(1)}^{(\infty)}(\Omega), \\ c_{tan}^{(\infty)}(\Omega) &= c_{(2)}^{(\infty)}(\Omega) + c_{(3)}^{(\infty)}(\Omega). \end{aligned}$$

In canonical way we extend these definitions to $c_{(i)}^{(p)}(\Omega)$, $0_i \leq p < \infty$ and $l^2(\Omega)$:

$$c_{(i)}^{(p)}(\Omega) = \{f \in c^{(p)}(\Omega) \,|\, O^{(k)} f = 0, \quad i \neq k\}$$

and

$$l_{(i)}^{(2)}(\Omega) = \overline{\{f \in c^{(\infty)}(\Omega) \,|\, O^{(k)} f = 0, \quad i \neq k\},}^{\|\cdot\|_{l^2(\Omega)}}$$

for $i = 1, 2, 3$. Obviously,

$$l_{(i)}^2(\Omega) = \overline{span\{y_{n,m}^{(i)}\}_{\substack{n \in \mathbb{N}_{0_i}; \\ m=1,\ldots,2n+1}}}^{\|\cdot\|_{l^2(\Omega)}}$$

and

$$l^2(\Omega) = \bigoplus_{i=1}^{3} l_{(i)}^2(\Omega).$$

The next theorem introduces the $l^2(\Omega)$-orthonormal system of vector spherical harmonics, derived from the $\mathcal{L}^2(\Omega)$-orthonormal system of scalar spherical harmonics.

Theorem 3.3.1. *Let* $\{Y_{n,m}\}_{\substack{n \in \mathbb{N}_0; \\ m=1,\ldots,2n+1}}$ *be an* $\mathcal{L}^2(\Omega)$*-orthonormal system of scalar spherical harmonics. Then the system*

$$y_{n,m}^{(i)} = (\mu_n^{(i)})^{-1/2} o^{(i)} Y_{n,m},$$

$n \in \mathbb{N}_{0_i}$; $m = 1, \ldots, 2n+1$; $i = 1, 2, 3$, *forms an* $l^2(\Omega)$*-orthonormal system of vector spherical harmonics when the normalization factor is chosen to be*

$$\mu_n^{(i)} = \left\|O^{(i)} o^{(i)} Y_{n,m}\right\|_{\mathcal{L}^2(\Omega)} = \begin{cases} 1, & \text{if} \quad i = 1, \\ n(n+1), & \text{if} \quad i = 2, 3. \end{cases}$$

We also mention that the system of vector spherical harmonics is closed and complete.

Definition 3.3.2. *A vector field* $h_n : \mathbb{R}^3 \to \mathbb{R}^3, n \in \mathbb{N}_0$, *is called a homogeneous harmonic vector polynomial of degree* n *if* $h_n \cdot \varepsilon^i$ *is a homogeneous harmonic polynomial of degree* $n \in \mathbb{N}_{0_i}$ *for every* $i = 1, 2, 3$.

We set

$$Harm_n(\mathbb{R}^3)\varepsilon^i = \{H_n\varepsilon^i | H_n \in Harm_n(\mathbb{R}^3)\},$$

and use this abbreviation to define the space of all homogeneous harmonic vector polynomials of degree n by

$$\bigoplus_{i=1}^{3} Harm_n(\mathbb{R}^3)\varepsilon^i.$$

In [14] it is shown that each type, 1 or 2, of vector spherical harmonics of degree n can be expressed as linear combinations of homogeneous harmonic vector polynomials of degree $n-1$ and $n+1$, restricted to the unit sphere, i.e.,

$$harm_n^{(i)} \subset \bigoplus_{j=1}^{3} Harm_{n-1}(\Omega)\varepsilon^j \oplus \bigoplus_{j=1}^{3} Harm_{n+1}(\Omega)\varepsilon^j, \quad i = 1, 2. \tag{3.3.1}$$

We find that, for type 3 vector spherical harmonics the relation

$$harm_n^{(3)} \subset \bigoplus_{j=1}^{3} Harm_n(\Omega)\varepsilon^j \tag{3.3.2}$$

holds true. Combining (3.3.1) and (3.3.2) we get

$$harm_n \subset \bigoplus_{m=n-1}^{n+1} \bigoplus_{i=1}^{3} Harm_n(\Omega)\varepsilon^i.$$

In analogy to the scalar case we know from [14] that

$$\bigoplus_{m=0}^{\infty} \bigoplus_{i=1}^{3} Harm_m\varepsilon^i$$

is dense in $c(\Omega)$ with respect to $\|\cdot\|_{c(\Omega)}$ and in $l^2(\Omega)$ with respect to $(.,.)_{l^2(\Omega)}$, so that we can state the following theorem:

Theorem 3.3.2. *Let* $\{y_{n,m}^{(i)}\}_{\substack{i=1,2,3;\, n\in\mathbb{N}_{0_i};\\ m=1,\ldots,2n+1}}$ *be an* $l^2(\Omega)$*-orthonormal system of vector spherical harmonics defined as in Theorem 3.3.1, then the following statements are valid:*

(i) The system $\{y_{n,m}^{(i)}\}_{\substack{i=1,2,3;\, n\in\mathbb{N}_{0_i};\\ m=1,\ldots,2n+1}}$ *of vector spherical harmonics is closed in* $c(\Omega)$ *with respect to* $\|\cdot\|_{c(\Omega)}$.

(ii) The system $\{y^{(i)}_{n,m}\}_{\substack{i=1,2,3;\, n\in\mathbb{N}_{0_i};\\ m=1,\ldots,2n+1}}$ *is complete in* $l^2(\Omega)$ *with respect to* $\|\cdot\|_{l^2(\Omega)}$.

From part (i) of this theorem we can deduce that any continuous vector field can be approximated arbitrarily close by finite linear combinations of vector spherical harmonics. Hence, by part (ii) we can represent any $f \in l^2(\Omega)$ by its Fourier series in terms of $\{y^{(i)}_{n,m}\}_{\substack{i=1,2,3;\, n\in\mathbb{N}_{0_i};\\ m=1,\ldots,2n+1}}$, i.e., we have

$$\lim_{N\to\infty} \left\| f - \sum_{i=1}^{3} \sum_{n=0_i}^{N} \sum_{m=1}^{2n+1} f^{(i)\wedge}(n,l) y^{(i)}_{n,m} \right\|_{l^{(2)}(\Omega)} = 0, \quad \text{for all } f \in l^2(\Omega)$$

with Fourier coefficients

$$f^{(i)\wedge}(n,l) = (f, y^{(i)}_{n,m})_{l^2(\Omega)} = \int_\Omega f(\xi)\cdot y^{(i)}_{n,m}(\xi) d\omega(\xi).$$

We may, of course, write

$$f = \sum_{i=1}^{3} f^{(i)},$$

where each $f^{(i)}$ is given by

$$f^{(i)} = \sum_{n=0_i}^{\infty} \sum_{m=1}^{2n+1} f^{(i)\wedge}(n,m) y^{(i)}_{n,m}.$$

The representation in terms of vector spherical harmonics enable us to model both the normal and the tangential part of any vector field $f \in l^2(\Omega)$. Thus, the Hilbert space $l^2(\Omega)$ can be split into three orthogonal subspaces that admit the following interpretation: the first subspace $l^2_{(1)}(\Omega)$ consists only of square-integrable normal fields, the second one contains only (surface-) curl-free tangential fields and the third one consists of (surface-) divergence-free tangential parts. Therefore, we can write:

$$\begin{aligned} l^2(\Omega) &= l^2_{nor}(\Omega) \oplus l^2_{tan}(\Omega), \\ l^2_{nor}(\Omega) &= l^2_{(1)}(\Omega), \\ l^2_{tan}(\Omega) &= l^2_{(2)}(\Omega) \oplus l^2_{(3)}(\Omega). \end{aligned}$$

We conclude this section with the vectorial analogon of the addition theorem. The first step to obtain the required theorem is to extend the definitions of the $o^{(i)}$−operators to vector fields. This can be done by referring to a sufficiently smooth vector field $f : \Omega \to \mathbb{R}^3$ on a sphere, admitting the representation

$$f(\xi) = \sum_{\nu=1}^{3} F_\nu(\xi)\varepsilon^\nu, \quad F_\nu(\xi) = f(\xi) \cdot \varepsilon^\nu,$$

where ε^ν are unit coordinate vectors. We define $o_\xi^{(i)} f(\xi)$ to be

$$o_\xi^{(i)} f(\xi) = \sum_{\nu=1}^{3} (o_\xi^{(i)} F_\nu(\xi)) \otimes \varepsilon^\nu, \quad i = 1, 2, 3.$$

Theorem 3.3.3. *Let* $\{y_{n,m}^{(i)}\}_{\substack{i=1,2,3;\, n\in\mathbb{N}_{0_i};\\ m=1,\ldots,2n+1}}$ *be an* $l^2(\Omega)$*-orthonormal basis . Then*

$$\sum_{m=1}^{2n+1} y_{n,m}^{(i)}(\xi) \otimes y_{n,m}^{(j)}(\eta) = \frac{2n+1}{4\pi} \mathbf{p}_n^{(i,j)}(\xi, \eta), \quad \xi, \eta \in \Omega,$$

with the (i, j)*-Legendre-tensor-field of degree* n*,* $i, j = 1, 2, 3$*, defined by*

$$\begin{aligned} &\mathbf{p}_n^{(i,j)}(\xi, \eta) : \Omega \times \Omega \to \mathbb{R}^3 \otimes \mathbb{R}^3, \\ &\mathbf{p}_n^{(i,j)}(\xi, \eta) = (\mu_n^{(k)})^{-1/2} (\mu_n^{(i)})^{-1/2} o_\xi^{(i)} o_\eta^{(j)} P_n(\xi \cdot \eta), \quad \xi, \eta \in \Omega. \end{aligned} \tag{3.3.3}$$

Detailed representations of the Legendre-tensor-fields can be found, e.g., in [12] and [14]. There exists an upper bound for the values $\mathbf{p}_n^{(i,j)}(\xi, \eta)$ for ξ, $\eta \in \Omega$, given by

Lemma 3.3.4. *Let* i*,* j*,* $l \in \{1, 2, 3\}$*. Then, for all* $\xi, \eta \in \Omega$*,*

$$|\mathbf{p}_n^{(i,j)}(\xi, \eta)\varepsilon^l| \leq 1.$$

For the Legendre tensors we find (see [12], [14],[33])

$$\begin{aligned} \mathbf{p}_n^{(1,1)}(\xi, \eta) &= P_n(\xi \cdot \eta)\xi \otimes \eta, \\ \mathbf{p}_n^{(1,2)}(\xi, \eta) &= \frac{1}{\sqrt{n(n+1)}} P_n'(\xi \cdot \eta)\xi \otimes (\xi - (\xi \cdot \eta)\eta), \\ \mathbf{p}_n^{(1,3)}(\xi, \eta) &= \frac{1}{\sqrt{n(n+1)}} P_n'(\xi \cdot \eta)\xi \otimes \eta \wedge \xi, \\ \mathbf{p}_n^{(2,1)}(\xi, \eta) &= \frac{1}{\sqrt{n(n+1)}} P_n'(\xi \cdot \eta)(\eta - (\xi \cdot \eta)\xi) \otimes \eta, \end{aligned}$$

$$
\begin{aligned}
\mathbf{p}_n^{(2,2)}(\xi,\eta) &= \frac{1}{\sqrt{2}}(P_n(\xi\cdot\eta)\mathbf{i}_{tan}(\xi),\\
\mathbf{p}_n^{(2,3)}(\xi,\eta) &= \frac{1}{n(n+1)}(P_n''(\xi\cdot\eta)(\eta-(\xi\cdot\eta)\xi)\otimes\eta\wedge\xi\\
&+ P_n'(\xi\cdot\eta)(-\mathbf{j}_{tan}(\eta)-\xi\otimes\eta\wedge\xi),\\
\mathbf{p}_n^{(3,1)}(\xi,\eta) &= \frac{1}{\sqrt{n(n+1)}}P_n'(\xi\cdot\eta)\xi\wedge\eta\otimes\xi,\\
\mathbf{p}_n^{(3,2)}(\xi,\eta) &= \frac{1}{n(n+1)}(P_n''(\xi\cdot\eta)\xi\wedge\eta\otimes(\xi-\xi\cdot\eta)\eta\\
&+ P_n'(\xi\cdot\eta)(\mathbf{j}_{tan}(\xi)-\xi\wedge\eta\otimes\eta),\\
\mathbf{p}_n^{(3,3)}(\xi,\eta) &= \frac{1}{\sqrt{2}}(P_n(\xi\cdot\eta)\mathbf{j}_{tan}(\xi),
\end{aligned}
$$

where the identity tensor $\mathbf{i}$ is defined by

$$\mathbf{i} = \sum_{i=1}^{3} \varepsilon^i \otimes \varepsilon^i,$$

and the surface identity tensor is given by

$$\mathbf{i}_{tan}(\xi) = \mathbf{i} - \xi\otimes\xi, \quad \xi\in\Omega,$$

and the surface rotation tensor is given by

$$\mathbf{j}_{tan}(\xi) = \xi\wedge\mathbf{i} = \sum_{i=1}^{3}(\xi\wedge\varepsilon^i)\otimes\varepsilon^i, \quad \xi\in\Omega.$$

The formulation of the vectorial addition theorem with Legendre-tensor-fields is a natural extension to the scalar case.
Introducing Legendre vectors $p_n^{(i)}:\Omega\times\Omega\to\mathbb{R}^3$, $i=1,2,3$, $n\in\mathbb{N}_{0_i}$, by

$$
\begin{aligned}
p_n^{(1)}(\xi,\eta) &= \xi P_n(\xi\cdot\eta)\\
p_n^{(2)}(\xi,\eta) &= \frac{1}{n(n+1)}(\eta-(\xi\cdot\eta)\xi)P_n'(\xi\cdot\eta),\\
p_n^{(3)}(\xi,\eta) &= \frac{1}{n(n+1)}(\xi\wedge\eta)P_n'(\xi\cdot\eta).
\end{aligned}
$$

we can formulate the following theorem:

Theorem 3.3.5. *Let* $\{Y_{n,m}\}_{\substack{n\in\mathbb{N}_0;\\ m=1,\ldots,2n+1}}$ *be a system of* $\mathcal{L}^2(\Omega)$*-orthonormal spherical harmonics and* $\{y_{n,m}^{(i)}\}_{\substack{i=1,2,3;\, n\in\mathbb{N}_{0_i};\\ m=1,\ldots,2n+1}}$ *the corresponding system of vector spherical harmonics. Then, for* $\xi,\eta\in\Omega$,

$$\sum_{m=1}^{2n+1} y_{n,m}^{(i)}(\xi)Y_{n,m}(\eta) = (\mu_n^{(i)})^{1/2}\frac{2n+1}{4\pi}p_n^{(i)}(\xi,\eta),$$

As we stated before the scalar spherical functions fulfill the Laplace equation: $\Delta_x r^n Y_n(\xi) = 0$, with $x = r\xi$, $\xi\in\Omega$, $r = |x|$. Easy calculations yield the following relations

$$\begin{aligned}
(\frac{1}{r^2}\frac{\partial}{\partial r}r^2\frac{\partial}{\partial r} + \frac{1}{r^2}\Delta_\xi^*)r^nY_n(\xi) &= 0,\\
(n(n-1)r^{n-2} + 2nr^{n-2} + r^{n-2}\Delta_\xi^*)Y_n(\xi) &= 0,\\
r^{n-2}(n(n+1)Y_n(\xi) + \Delta_\xi^*)Y_n(\xi) &= 0,\\
(\Delta_\xi^* + n(n+1))Y_n(\xi) &= 0.
\end{aligned}$$

The last equation indicates that the scalar spherical harmonics are eigenfunctions of the Beltrami differential operator to the eigenvalues $-n(n+1)$. The vector spherical harmonics for $i = 1, 2$, i.e., $\{y_{n,m}^{(i)}\}_{\substack{i=1,2;\, n\in\mathbb{N}_{0_i};\\ m=1,\ldots,2n+1}}$ do not share this property. They are not eigenfunctions of the scalar Beltrami operator. This is stated in the next theorem.

Theorem 3.3.6. *Let* $Y_n \in Harm_n$ *be a spherical harmonic of degree* n. *Then we have*

$$\Delta^* o^{(1)}Y_n = (-n(n+1)-2)o^{(1)}Y_n + 2o^{(2)}Y_n, \tag{3.3.4}$$

$$\Delta^* o^{(2)}Y_n = 2n(n+1)o^{(1)}Y_n - n(n+1)o^{(2)}Y_n, \tag{3.3.5}$$

$$\Delta^* o^{(3)}Y_n = -n(n+1)o^{(3)}Y_n, \tag{3.3.6}$$

where the (scalar) Beltrami operator Δ^* *for a function* $f\in c^{(2)}(\Omega)$ *of the form*

$$f(\xi) = \sum_{i=1}^{3}\varepsilon^i F_i(\xi),$$

is defined by

$$\Delta_\xi^* f(\xi) = \sum_{i=1}^{3}\varepsilon^i\Delta_\xi^* F_i(\xi), \quad \xi\in\Omega.$$

The proof of Theorem 3.3.6 can be found in [14]. We deduce that the vector spherical harmonics $\{y_{n,m}^{(i)}\}_{\substack{i=1,2,3;\, n\in\mathbb{N}_{0_i};\\ m=1,\ldots,2n+1}}$ do not fulfill the Laplace equation.

(Edmonds-)Vector Spherical Harmonics $u_{n,m}^{(i)}$

Now we are interested in the construction of vector fields which fulfill the Laplace equation. Therefore, we introduce another set of vector spherical harmonics which are constructed by operators $k_n^{(i)}$, $i = 1,2,3$, as proposed in quantum mechanics [5].
Suppose that $H_n \in Harm_n(\mathbb{R}^n)$. Considering the vector field ∇H_n, we realize that it is a homogeneous harmonic vector polynomial of degree $n-1$. Furthermore, it is not hard to observe that $x \to x \wedge \nabla_x H_n(x)$ for $x \in \mathbb{R}^3$, represents a homogeneous harmonic vector polynomial of degree n. The function $x \to xH_n(x)$, $x \in \mathbb{R}^3$, is, in general, not harmonic. Therefore, we go over to the function $x \to ((2n+1)x - |x|^2\nabla_x)H_n(x)$, which turns out to be a homogeneous harmonic vector polynomial of degree $n+1$. This perspective motivates the following definition.

Definition 3.3.3. *For $n \in \mathbb{N}_0$ and $x \in \mathbb{R}^3$, F sufficiently smooth, we define the operators $k_n^{(i)}$, $i = 1,2,3$, by*

$$k_n^{(1)}F(x) = ((2n+1)x - |x|^2\nabla_x)F(x), \tag{3.3.7}$$

$$k_n^{(2)}F(x) = \nabla_x F(x), \tag{3.3.8}$$

$$k_n^{(3)}F(x) = x \wedge \nabla_x F(x). \tag{3.3.9}$$

The definition of the operators $k_n^{(i)}$ leads to the following lemma.

Lemma 3.3.7. *Let $H_n \in Harm_n(\mathbb{R}^3)$, $n \in \mathbb{N}_0$. Then $k_n^{(i)}H_n$ is a homogeneous harmonic vector polynomial of degree $deg^{(i)}(n)$, where we use the abbreviation*

$$deg^{(i)}(n) = \begin{cases} n+1, & i=1, \\ n-1, & i=2, \\ n, & i=3. \end{cases}$$

If $deg^{(i)}(n) < 0$ then $k_n^{(i)}H_n = 0$.

Applying the operators $k_n^{(i)}$ on $H_n \in Harm_n(\mathbb{R}^3)$, with $H_n(x) = r^nY_n(\xi)$ yields

$$\mathrm{h}_n^{(1)}(x) = k_n^{(1)}r^nY_n(\xi) = (n+1)r^{n+1}o^{(1)}Y_n(\xi) - r^{n+1}o^{(2)}Y_n(\xi),$$
$$\mathrm{h}_n^{(2)}(x) = k_n^{(2)}r^nY_n(\xi) = nr^{n-1}o^{(1)}Y_n(\xi) - r^{n-1}o^{(2)}Y_n(\xi),$$
$$\mathrm{h}_n^{(3)}(x) = k_n^{(3)}r^nY_n(\xi) = r^no^{(3)}Y_n(\xi),$$

where $\{\mathrm{h}_n^{(i)}\}_{i=1,2,3;\, n\in\mathbb{N}_{0_i}}$ represents a set of vector fields. Therefore, the restrictions of $x \mapsto \mathrm{h}_n^{(i)}(x) = k_n^{(i)} r^n Y_n(\xi)$, $x = r\xi$, $r = |x|$, to the unit sphere Ω can be written as linear combinations of vector spherical harmonics $o^{(i)} Y_n$.
In the sequel we identify $k_n^{(i)} Y_n$ with

$$k_n^{(i)} Y_n(\xi) = \mathrm{h}_n^{(i)}(x)|_{|x|=1} = k_n^{(i)} H_n(x)|_{r=1},$$

with $H_n(x) = r^n Y_n(\xi)$, $x = r\xi$, or, in more detail,

$$k_n^{(1)} Y_n(\xi) = (n+1) o^{(1)} Y_n(\xi) - o^{(2)} Y_n(\xi), \tag{3.3.10}$$

$$k_n^{(2)} Y_n(\xi) = n o^{(1)} Y_n(\xi) + o^{(2)} Y_n(\xi), \tag{3.3.11}$$

$$k_n^{(3)} Y_n(\xi) = o^{(3)} Y_n(\xi). \tag{3.3.12}$$

Let $G \in Harm_n$ then the adjoint operators satisfying

$$(k_n^{(i)} G, f)_{l^2(\Omega)} = (G, K_n^{(i)} f)_{\mathcal{L}^2(\Omega)}, \quad i = 1, 2, 3,$$

where the adjoint operators $K_n^{(i)}$ to $k_n^{(i)}$ are given by

$$\begin{aligned} K_n^{(1)} f &= (n+1) O^{(1)} f - O^{(2)} f, \\ K_n^{(2)} f &= n O^{(1)} f + O^{(2)} f, \\ K_n^{(3)} f &= O^{(3)} f, \end{aligned}$$

with $f \in harm_n$.
The scale factor for the vector spherical harmonics is defined in the next lemma.

Lemma 3.3.8. *For an orthonormal system of spherical harmonics* $\{Y_{n,m}\}_{\substack{n\in\mathbb{N}_0;\\ m=1,\ldots,2n+1}}$ *the following equations hold true:*

$$K_n^{(i)} k_n^{(j)} Y_{n,m}(\xi) = \delta_{i,j} \nu_n^{(i)} Y_{n,m}(\xi), \quad i, j = 1, 2, 3, \tag{3.3.13}$$

where the constants $\nu_n^{(i)}$ *are defined by*

$$\nu_n^{(i)} = \|K_n^{(i)} k_n^{(i)} Y_{n,m}\|_{\mathcal{L}^2(\Omega)},$$

i.e.,

$$\nu_n^{(1)} = (n+1)(2n+1), \tag{3.3.14}$$

$$\nu_n^{(2)} = n(2n+1), \tag{3.3.15}$$

$$\nu_n^{(3)} = n(n+1). \tag{3.3.16}$$

In other words, we are led to the following set of vector spherical harmonics (note that we base our considerations on the same approach as introduced by [5]):

Definition 3.3.4. *Any vector field*

$$u_n^{(i)} = k_n^{(i)} Y_n, \quad i = 1,2,3, \; n \in \mathbb{N}_{0_i}, \quad Y_n \in Harm_n(\Omega),$$

is called an (Edmonds-)vector spherical harmonic of degree n and type i.

Lemma 3.3.9. *Let* $\{Y_{n,m}\}_{\substack{n\in\mathbb{N}_0;\\ m=1,\ldots,2n+1}}$ *be an* $\mathcal{L}^2(\Omega)$*-orthonormal system of scalar spherical harmonics. Then the vector fields given by*

$$u_{n,m}^{(i)} = (\nu_n^{(i)})^{-1/2} k_n^{(i)} Y_{n,m},$$

$n \in \mathbb{N}_{0_i}$; $m = 1,\ldots,2n+1$; $i = 1,2,3$, *form an* $l^2(\Omega)$*-orthonormal set of vector spherical harmonics with the normalization coefficients as defined in (3.3.14)-(3.3.16).*

By inverting (3.3.10)-(3.3.12) we obtain the following equations for $\xi \in \Omega$:

$$o^{(1)} Y_{n,m}(\xi) = \frac{1}{2n+1}\left(k_n^{(1)} Y_{n,m}(\xi) + k_n^{(2)} Y_{n,m}(\xi)\right), \tag{3.3.17}$$

$$o^{(2)} Y_{n,m}(\xi) = \frac{1}{2n+1}\left(-n k_n^{(1)} Y_{n,m}(\xi) + (n+1) k_n^{(2)} Y_{n,m}(\xi)\right), \tag{3.3.18}$$

$$o^{(3)} Y_{n,m}(\xi) = k_n^{(3)} Y_{n,m}(\xi). \tag{3.3.19}$$

This provides a relation between the system $\{y_{n,m}^{(i)}\}_{\substack{i=1,2,3;n\in\mathbb{N}_{0_i};\\ m=1,\ldots,2n+1}}$ and $\{u_{n,m}^{(i)}\}_{\substack{i=1,2,3;n\in\mathbb{N}_{0_i};\\ m=1,\ldots,2n+1}}$ of vector spherical harmonics. More explicitly, these systems are connected in the following way

$$\begin{aligned} u_{n,m}^{(1)} &= \sqrt{\frac{n+1}{2n+1}}\, y_{n,m}^{(1)} - \sqrt{\frac{n}{2n+1}}\, y_{n,m}^{(2)}, \\ u_{n,m}^{(2)} &= \sqrt{\frac{n}{2n+1}}\, y_{n,m}^{(1)} + \sqrt{\frac{n+1}{2n+1}}\, y_{n,m}^{(2)}, \\ u_{n,m}^{(3)} &= y_{n,m}^{(3)}. \end{aligned} \tag{3.3.20}$$

Conversely,

$$\begin{aligned} y_{n,m}^{(1)} &= \sqrt{\frac{n+1}{2n+1}}\, u_{n,m}^{(1)} + \sqrt{\frac{n}{2n+1}}\, u_{n,m}^{(2)}, \\ y_{n,m}^{(2)} &= -\sqrt{\frac{n}{2n+1}}\, u_{n,m}^{(1)} + \sqrt{\frac{n+1}{2n+1}}\, u_{n,m}^{(2)} \\ y_{n,m}^{(3)} &= u_{n,m}^{(3)}. \end{aligned}$$

An immediate consequence of the construction of the new system of vector spherical harmonics is the following corollary (see [14]).

Theorem 3.3.10. *Let the system of vector spherical harmonics* $\{u_{n,m}\}_{\substack{i=1,2,3;n\in\mathbb{N}_{0_i};\\ m=1,\dots,2n+1}}$ *be defined as in Lemma 3.3.9. Then the following statements are valid*

(i) The above system of vector spherical harmonics is closed in $c(\Omega)$ *with respect to* $\|\cdot\|_{c(\Omega)}$.

(ii) The above system of vector spherical harmonics is complete in $l^2(\Omega)$ *with respect to* $\|\cdot\|_{l^2(\Omega)}$.

From part (i) of Theorem 3.3.10 we can deduce that any continuous harmonic vector field can be approximated arbitrarily close by finite linear combinations of vector spherical harmonics $\{u^{(i)}_{n,m}\}_{\substack{i=1,2,3;n\in\mathbb{N}_{0_i};\\ m=1,\dots,2n+1}}$. Hence, by part (ii) we can represent any function $f \in l^2(\Omega)$ by its Fourier series in terms of $\{u^{(i)}_{n,m}\}_{\substack{i=1,2,3;n\in\mathbb{N}_{0_i};\\ m=1,\dots,2n+1}}$, i.e.,

$$\lim_{N\to\infty}\left\| f - \sum_{i=1}^{3}\sum_{n=0_i}^{N}\sum_{m=1}^{2n+1} f^{(i)\wedge}(n,m)u^{(i)}_{n,m}\right\|_{l^{(2)}(\Omega)} = 0,$$

for all $f \in l^2(\Omega)$ with Fourier coefficients

$$f^{(i)\wedge}(n,m) = (f, u^{(i)}_{n,m})_{l^2(\Omega)} = \int_\Omega f(\xi)\cdot u^{(i)}_{n,m}(\xi)\,d\omega(\xi).$$

We may, of course, write

$$f = \sum_{i=1}^{3} f^{(i)},$$

where each $f^{(i)}$ is given by

$$f^{(i)} = \sum_{n=0_i}^{\infty}\sum_{m=1}^{2n+1} f^{(i)\wedge}(n,m)u^{(i)}_{n,m}.$$

Theorem 3.3.11. *Let* $\{Y_{n,m}\}_{\substack{n\in\mathbb{N}_0;\\ m=1,\dots,2n+1}}$ *be an* $\mathcal{L}^2(\Omega)$*-orthonormal set of scalar spherical harmonics. Then the set* $\{u^{(i)}_{n,m}\}_{\substack{i=1,2,3;n\in\mathbb{N}_{0_i};\\ m=1,\dots,2n+1}}$ *defined in (3.3.9) forms an* $l^2(\Omega)$*-orthonormal* *set of vector spherical harmonics which is closed in* $c(\Omega)$ *with respect to* $\|\cdot\|_{c(\Omega)}$ *and complete in* $l^2(\Omega)$ *with respect to* $(\cdot,\cdot)_{l^2(\Omega)}$. *Furthermore, for all* $\xi \in \Omega$,

$$\begin{aligned}
\Delta^*_\xi u^{(1)}_{n,m}(\xi) &= -(n+1)(n+2)u^{(1)}_{n,m}(\xi),\\
\Delta^*_\xi u^{(2)}_{n,m}(\xi) &= -n(n-1)u^{(2)}_{n,m}(\xi),\\
\Delta^*_\xi u^{(3)}_{n,m}(\xi) &= -n(n+1)u^{(3)}_{n,m}(\xi).
\end{aligned}$$

In other words,

$$\begin{aligned}
\Delta^*_\xi u^{(1)}_{n-1,m}(\xi) &= -n(n+1)u^{(1)}_{n-1,m}(\xi),\\
\Delta^*_\xi u^{(2)}_{n+1,m}(\xi) &= -n(n+1)u^{(2)}_{n+1,m}(\xi),\\
\Delta^*_\xi u^{(3)}_{n,m}(\xi) &= -n(n+1)u^{(3)}_{n,m}(\xi).
\end{aligned}$$

On the one hand, each member of the (Edmonds-)system $\{u^{(i)}_{n,m}\}_{\substack{i=1,2,3;n\in\mathbb{N}_{0_i};\\ m=1,\ldots,2n+1}}$ is, by definition, not decomposable into normal and tangential parts, but on the other hand it is a set of eigenfunctions of the Beltrami operator. This property will enable us in Chapter 4 to introduce a set of vector outer harmonics which fulfill the Laplace equation outside a sphere Ω_R.

Further, we can show that for any $f \in l^2(\Omega)$ Parseval's identity holds true, thus we have

$$(f,f)_{l^2(\Omega)} = \sum_{i=1}^{3}\sum_{n=0_i}^{\infty}\sum_{m=1}^{2n+1}(f,u^{(i)}_{n,m})^2_{l^2(\Omega)}.$$

Analogously to the function spaces for scalar and vector spherical harmonics we define for the system $\{u^{(i)}_{n,m}\}_{\substack{i=1,2,3;n\in\mathbb{N}_{0_i};\\ m=1,\ldots,2n+1}}$ the function spaces

$$\widetilde{harm}^{(i)}_n = span\{u^{(i)}_{n,m}\}_{\substack{n\in\mathbb{N}_{0_i};\\ m=1,\ldots,2n+1}},$$

for $i=1,2,3$, which fulfill

$$\begin{aligned}
harm^{(1)}_0 &= \widetilde{harm}^{(1)}_0,\\
harm^{(1)}_n \oplus harm^{(2)}_n &= \widetilde{harm}^{(1)}_n \oplus \widetilde{harm}^{(2)}_n, \quad n\in\mathbb{N},\\
harm^{(3)}_n &= \widetilde{harm}^{(3)}_n, \quad n\in\mathbb{N}.
\end{aligned}$$

Thus, we have

$$\begin{aligned}
harm_0 &= \widetilde{harm}^{(1)}_0,\\
harm_n &= \bigoplus_{i=1}^{3}\widetilde{harm}^{(i)}_n, \quad n\in\mathbb{N}.
\end{aligned}$$

The new set of vector spherical harmonics is characterized by the property that the Laplace equation is fulfilled. Thus, as shown by the next lemma, homogeneous harmonic vector polynomials can be composed by vector spherical harmonics of different degrees.

Lemma 3.3.12. *Let for $k = 1, 2, 3$, $n \in \mathbb{N}_{0_i}$, $\varepsilon^k H_n$ be a homogeneous harmonic vector polynomial. Then*

$$\varepsilon^k H_n|_\Omega = u^{(1)}_{n-1} + u^{(2)}_{n+1} + u^{(3)}_n,$$

with

$$\begin{aligned} u^{(1)}_{n-1} &= k^{(1)}_{n-1} Y_{n-1}, \\ u^{(2)}_{n+1} &= k^{(2)}_{n+1} Y_{n+1}, \\ u^{(3)}_n &= k^{(3)}_n Y_n. \end{aligned}$$

Proof. As the homogeneous harmonic vector polynomial $\varepsilon^k H_n$ is an element of $l^2(\Omega)$, we get with Theorem 3.3.10

$$\varepsilon^k H_n|_\Omega = \sum_{i=1}^{3} \sum_{p=0_i}^{\infty} \sum_{q=1}^{2p+1} a^{(i)}_{p,q} u^{(i)}_{p,q},$$

where $\{u^{(i)}_{p,q}\}_{\substack{i=1,2,3; p\in\mathbb{N}_{0_i};\\ q=1,\ldots,2p+1}}$ is an orthonormal system of vector spherical harmonics. Furthermore, because of Lemma 3.3.7 the following equations hold true.

$$u^{(1)}_{n,m} = \sum_{j=1}^{3} c^{(1)}_{j,m} \varepsilon^j Y^j_{n+1}, \quad Y^j_{n+1} \in Harm_{n+1}, \tag{3.3.21}$$

$$u^{(2)}_{n,m} = \sum_{j=1}^{3} c^{(2)}_{j,m} \varepsilon^j Y^j_{n-1}, \quad Y^j_{n-1} \in Harm_{n-1}, \tag{3.3.22}$$

$$u^{(3)}_{n,m} = \sum_{j=1}^{3} c^{(3)}_{j,m} \varepsilon^j Y^j_{n}, \quad Y^j_{n} \in Harm_{n}. \tag{3.3.23}$$

The system $\{u^{(i)}_{n,m}\}_{\substack{i=1,2,3; n\in\mathbb{N}_{0_i};\\ m=1,\ldots,2n+1}}$ builds an orthonormal basis in $l^2(\Omega)$, thus we finally get

$$a^{(1)}_{p,q} = \begin{cases} 0, & n-1 \neq p, \\ C^{(1)}_{k,p,q}, & n-1 = p, \end{cases}$$

$$a^{(2)}_{p,q} = \begin{cases} 0, & n+1 \neq p, \\ C^{(2)}_{k,p,q}, & n+1 = p, \end{cases}$$

$$a^{(3)}_{p,q} = \begin{cases} 0, & n \neq p, \\ C^{(3)}_{k,p,q}, & n = p, \end{cases}$$

with constants $C^{(i)}_{k,p,q} \in \mathbb{R}$.
This yields the desired result. □

In a similar way as for the vector spherical harmonics $\{y^{(i)}_{n,m}\}_{\substack{i=1,2,3;n\in\mathbb{N}_{0_i};\\ m=1,\ldots,2n+1}}$ an addition theorem can be formulated. For that reason we first define the Legendre tensors based on the vector fields $\{u^{(i)}_{n,m}\}_{\substack{i=1,2,3;n\in\mathbb{N}_{0_i};\\ m=1,\ldots,2n+1}}$.

Definition 3.3.5. *The (i,j)-Legendre tensor field $\tilde{\mathbf{p}}^{(i,j)}_n : \Omega \times \Omega \to \mathbb{R}^3 \otimes \mathbb{R}^3$, $i,j = 1,2,3$, of degree n is given by*

$$\tilde{\mathbf{p}}^{(i,j)}_n(\xi,\eta) = (\nu^{(i)}_n)^{-1/2}(\nu^{(j)}_n)^{-1/2}(k^{(i)}_n)_\xi (k^{(j)}_n)_\eta P_n(\xi\cdot\eta), \quad \xi,\eta \in \Omega, \tag{3.3.24}$$

where

$$(k^{(i)}_n)_\xi f(\xi) = \sum_{l=1}^{3} (k^{(i)}_n F_l(\xi)) \otimes \varepsilon^l, \quad i = 1,2,3,$$

for any sufficiently smooth vector field $f : \Omega \to \mathbb{R}^3$ of the form

$$f(\xi) = \sum_{l=1}^{3} F_l(\xi)\varepsilon^l.$$

Next we introduce the addition theorem for vector spherical harmonics $\{u^{(i)}_{n,m}\}_{\substack{i=1,2,3;n\in\mathbb{N}_{0_i};\\ m=1,\ldots,2n+1}}$.

Theorem 3.3.13. *Let $\{u^{(i)}_{n,m}\}_{\substack{i=1,2,3;n\in\mathbb{N}_{0_i};\\ m=1,\ldots,2n+1}}$ be an $l^2(\Omega)$-orthonormal basis as defined in Lemma 3.3.9. Then*

$$\sum_{m=1}^{2n+1} u^{(i)}_{n,m}(\xi) \otimes u^{(k)}_{n,m}(\eta) = \frac{2n+1}{4\pi}\tilde{\mathbf{p}}^{(i,k)}_n(\xi,\eta)$$

holds for $i,k = 1,2,3$.

Proof. Observing the definitions of the Legendre tensors, given in (3.3.3) for $\{y_{n,m}^{(i)}\}_{\substack{i=1,2,3;n\in\mathbb{N}_{0_i};\\ m=1,\dots,2n+1}}$ and the addition Theorem 3.3.13 we get the desired result. □

The Legendre tensors $\tilde{\mathbf{p}}_n^{(i,k)}$ (given by (3.3.24)) have the same upper bound as the Legendre tensors of $\{y_{n,m}^{(i)}\}_{\substack{i=1,2,3;n\in\mathbb{N}_{0_i};\\ m=1,\dots,2n+1}}$ (given in Lemma 3.3.4).
The connection between the Legendre tensors $\tilde{\mathbf{p}}_n^{(i,j)}$ and the Legendre tensors $\mathbf{p}_n^{(i,j)}$ (see (3.3.3)) is given by the following lemma [33].

Lemma 3.3.14. *Let the tensor fields $\tilde{\mathbf{p}}_n^{(i,j)} : \Omega \times \Omega \to \mathbb{R}^3 \otimes \mathbb{R}^3$, $i, j = 1, 2, 3$, be defined by (3.3.24). Then we have*

$$
\begin{aligned}
\tilde{\mathbf{p}}_n^{(1,1)} &= (n+1)^2 (c_n^{1,1})^2 \mathbf{p}_n^{(1,1)} - (n+1) c_n^{1,1} c_n^{2,1} (\mathbf{p}_n^{(1,2)} + \mathbf{p}_n^{(2,1)}) + (c_n^{2,1})^2 \mathbf{p}_n^{(2,2)}, \\
\tilde{\mathbf{p}}_n^{(1,2)} &= n(n+1) c_n^{1,1} c_n^{1,2} \mathbf{p}_n^{(1,1)} + (n+1) c_n^{1,1} c_n^{2,2} \mathbf{p}_n^{(1,2)} - n c_n^{1,1} c_n^{2,2} \mathbf{p}_n^{(2,1)} - c_n^{2,1} c_n^{2,2} \mathbf{p}_n^{(2,2)}, \\
\tilde{\mathbf{p}}_n^{(1,3)} &= (n+1) c_n^{1,1} c_n^{3,3} \mathbf{p}_n^{(1,3)} - c_n^{2,1} c_n^{3,3} \mathbf{p}_n^{(2,3)}, \\
\tilde{\mathbf{p}}_n^{(2,1)} &= n(n+1) c_n^{1,2} c_n^{1,1} \mathbf{p}_n^{(1,1)} + (n+1) c_n^{2,2} c_n^{1,1} \mathbf{p}_n^{(2,1)} - n c_n^{1,2} c_n^{2,1} \mathbf{p}_n^{(1,2)} - c_n^{2,2} c_n^{2,1} \mathbf{p}_n^{(2,2)}, \\
\tilde{\mathbf{p}}_n^{(2,2)} &= n^2 (c_n^{1,2})^2 \mathbf{p}_n^{(1,1)} + n c_n^{1,2} c_n^{2,2} (\mathbf{p}_n^{(1,2)} + \mathbf{p}_n^{(2,1)}) + (c_n^{2,2})^2 \mathbf{p}_n^{(2,2)}, \\
\tilde{\mathbf{p}}_n^{(2,3)} &= n c_n^{1,2} c_n^{3,3} \mathbf{p}_n^{(1,3)} + c_n^{2,2} c_n^{3,3} \mathbf{p}_n^{(2,3)}, \\
\tilde{\mathbf{p}}_n^{(3,1)} &= (n+1) c_n^{1,1} c_n^{3,1} \mathbf{p}_n^{(3,1)} - c_n^{3,3} c_n^{2,1} \mathbf{p}_n^{(3,2)}, \\
\tilde{\mathbf{p}}_n^{(3,2)} &= n c_n^{3,3} c_n^{1,2} \mathbf{p}_n^{(3,1)} + c_n^{3,3} c_n^{1,2} \mathbf{p}_n^{(3,2)}, \\
\tilde{\mathbf{p}}_n^{(1,3)} &= (c_n^{3,3})^2 \mathbf{p}_n^{(3,3)},
\end{aligned}
$$

where the constants $c_n^{i,k}$ are given by

$$c_n^{i,k} = \sqrt{\frac{\mu_n^{(i)}}{\nu_n^{(k)}}}.$$

We introduce Legendre vectors $\tilde{p}_n^{(i)}$ which are based on the system $\{u_{n,m}^{(i)}\}_{\substack{i=1,2,3;n\in\mathbb{N}_{0_i};\\ m=1,\dots,2n+1}}$ and which allow us to constitute the addition theorem in an alternative way.

Definition 3.3.6. *The i-Legendre vector field for the system $\{u_{n,m}^{(i)}\}_{\substack{i=1,2,3;n\in\mathbb{N}_{0_i};\\ m=1,\dots,2n+1}}$ of degree n, $\tilde{p}_n^{(i)} : \Omega \times \Omega \to \mathbb{R}^3$, $i = 1, 2, 3$, is given by*

$$\tilde{p}_n^{(i)}(\xi, \eta) = (\nu_n^{(i)})^{-1/2} (k_n^{(i)})_\xi \, P_n(\xi \cdot \eta), \quad \xi, \eta \in \Omega.$$

The connection between the Legendre vectors for $\{y^{(i)}_{n,m}\}_{\substack{i=1,2,3;n\in\mathbb{N}_{0_i};\\ m=1,\ldots,2n+1}}$ and the Legendre vectors for $\{u^{(i)}_{n,m}\}_{\substack{i=1,2,3;n\in\mathbb{N}_{0_i};\\ m=1,\ldots,2n+1}}$ is given by the following lemma.

Lemma 3.3.15. *Let the Legendre vectors* $\tilde{p}^{(i)}_n : \Omega \times \Omega \to \mathbb{R}^3$, $i = 1, 2, 3$, *be defined as in Definition 3.3.6. Then we have*

$$\begin{aligned}
\tilde{p}^{(1)}_n(\xi,\eta) &= (n+1)c^{1,1}_n p^{(1)}_n - c^{2,1}_n p^{(2)}_n, \\
\tilde{p}^{(2)}_n(\xi,\eta) &= n c^{1,2}_n p^{(1)}_n - c^{2,2}_n p^{(2)}_n, \\
\tilde{p}^{(3)}_n(\xi,\eta) &= c^{3,3}_n p^{(1)}_n,
\end{aligned}$$

where the constants $c^{(i,k)}_n$, $i, k = 1, 2, 3$, *are given by* $c^{i,k}_n = \sqrt{\frac{\mu^{(i)}_n}{\nu^{(k)}_n}}$.

With the help of the Legendre vectors for $\{u^{(i)}_{n,m}\}_{\substack{i=1,2,3;n\in\mathbb{N}_{0_i};\\ m=1,\ldots,2n+1}}$ the addition theorem then reads as follows.

Theorem 3.3.16. *Let* $\{Y_{n,m}\}_{\substack{n\in\mathbb{N}_0;\\ m=1,\ldots,2n+1}}$ *be an* $\mathcal{L}^2(\Omega)$*-orthonormal basis of* $Harm_n$ *and* $u^{(i)}_{n,m} = (\nu^{(i)}_n)^{-1/2} k^{(i)}_n Y_{n,m}$, $n \in \mathbb{N}_{0_i}$; $m = 1, \ldots, 2n+1$; $i = 1, 2, 3$. *Then*

$$\sum_{m=1}^{2n+1} u^{(i)}_{n,m}(\xi) Y_{n,m}(\eta) = \frac{2n+1}{4\pi} \tilde{p}^{(i)}_n(\xi,\eta), \quad \xi, \eta \in \Omega,$$

holds for $i = 1, 2, 3$.

Observing relations (see [24]):

$$\begin{aligned}
\nabla_x \cdot u^{(2)}_{n,m}(x) = 0, &\quad \nabla_x \wedge u^{(2)}_{n,m}(x) = 0, \\
x \cdot u^{(3)}_{n,m}(x) = 0, &\quad \nabla_x \cdot u^{(3)}_{n,m}(x) = 0,
\end{aligned}$$

we see that $u^{(2)}_{n,m}(x)$ is a poloidal field whereby $u^{(3)}_{n,m}(x)$ is a toroidal field.

Chapter 4

Scalar and Vector Outer Harmonics

In this chapter, first, we extend the theory of spherical harmonics from the unit sphere to a sphere with radius $R \in \mathbb{R}$, $R > 0$. Second, we introduce scalar outer harmonics (see [18], [33]) and vector outer harmonics (see [14], [33]). Further, we investigate the closure property of vector outer harmonics with respect to a regular surface Σ. Then two algorithms for the exact generation for homogeneous harmonic polynomials are presented followed by orthonormalization procedures. From that the calculation of scalar spherical harmonics and scalar outer harmonics are derived. We close this chapter with the exact calculation of vector spherical harmonics and vector outer harmonics which builds the fundamentals for the numerical realization of the approximation methods described in Chapter 5. Referring to Figure 1.0.1 we deal in this chapter with step (2) and (3).

4.1 Extension to the Sphere Ω_R

Up to now, we have only considered a sphere Ω around the origin with radius $R = 1$. As mentioned before Ω_R is a sphere of radius $R \in \mathbb{R}$, $R > 0$, where we define the inner space by $\Omega_R^{int} = \{x \in \mathbb{R} \,|\, |x| < R\}$ and the outer space by $\Omega_R^{ext} = \{x \in \mathbb{R} \,|\, |x| > R\}$.
By virtue of the isomorphism $\Omega \ni \xi \mapsto R\xi \in \Omega_R$, functions given on Ω can be understood to operate on Ω_R, and vice versa. We define the Hilbert space $(\mathcal{L}^2(\Omega_R), (\cdot,\cdot)_{\mathcal{L}^2(\Omega_R)})$ by using the scalar product

$$(F,G)_{\mathcal{L}^2(\Omega_R)} = \int_{\Omega_R} F(R\xi)G(R\xi)d\omega_R(\xi).$$

With the relationship $\xi \leftrightarrow R\xi$, every function $F : \Omega \to \mathbb{R}$ can be viewed as a function on Ω_R by defining $\tilde{F} : \Omega_R \to \mathbb{R}$ with $\tilde{F}(R\xi) = F(\xi)$, where as before Ω_R defines a sphere with radius $R \in \mathbb{R}$, $R > 0$. In order to transfer the differential operators to spheres with arbitrary radius, we abbreviate the new operators by giving them a superscript R and define (see, e.g., [18])

$$\nabla^{*;R} = \frac{1}{R}\nabla^*, \quad L^{*;R} = \frac{1}{R}L^*, \quad \Delta_\xi^{*;R} = \frac{1}{R^2}\Delta_\xi^*.$$

Thus we get an orthonormal basis in $\mathcal{L}^2(\Omega_R)$. Consequently, every function $F \in \mathcal{L}^2(\Omega_R)$ can be written as a Fourier series

$$F = \sum_{n=0}^{\infty}\sum_{m=1}^{2n+1} F^{\wedge_R}(n,m) Y_{n,m}^R,$$

(in $\mathcal{L}^2(\Omega_R)$-sense) with Fourier coefficients

$$F^{\wedge_R}(n,m) = (F, Y_{n,m}^R)_{\mathcal{L}^2(\Omega_R)}$$

and

$$Y_{n,m}^R(x) = \frac{1}{R} Y_{n,m}\left(\frac{x}{|x|}\right), \quad x \in \Omega_R. \tag{4.1.1}$$

Analogously, we are able to define the system $\{y_{n,m}^{(i),R}\}_{\substack{i=1,2,3;\, n\in\mathbb{N}_{0_i};\\ m=1,\ldots,2n+1}}$ of vector spherical harmonics for $\{y_{n,m}^{(i)}\}_{\substack{i=1,2,3; n\in\mathbb{N}_{0_i};\\ m=1,\ldots,2n+1}}$ for an arbitrary sphere Ω_R by

$$y_{n,m}^{(i),R}(x) = \frac{1}{R} y_{n,m}^{(i)}\left(\frac{x}{|x|}\right), \quad x \in \Omega_R. \tag{4.1.2}$$

As the system $\{y_{n,m}^{(i)}\}_{\substack{i=1,2,3;\, n\in\mathbb{N}_{0_i};\\ m=1,\ldots,2n+1}}$ constitutes an $l^2_{(i)}(\Omega)$-orthonormal basis for $i = 1,2,3$, the system $\{y_{n,m}^{(i),R}\}_{\substack{i=1,2,3;\, n\in\mathbb{N}_{0_i};\\ m=1,\ldots,2n+1}}$ also forms an orthonormal basis for $i = 1,2,3$ in $l^2_{(i)}(\Omega_R)$.

The projection operators given in (2.0.7) can be analogously extended to p^R_{nor} and p^R_{tan} for the sphere Ω_R, and, in addition, the function spaces $c(\Omega_R)$ and $l^2_{(i)}(\Omega_R)$ can be introduced in an analogous way to (2.0.8) - (2.0.11).

Every vector field $f^{(i)} \in l^2_{(i)}(\Omega_R)$ can be expanded in terms of vector spherical harmonics for Ω_R as follows:

$$f^{(i)} = \sum_{n=0_i}^{\infty}\sum_{m=1}^{2n+1} f^{(i)\wedge_R}(n,m) y_{n,m}^{(i),R}, \quad i = 1,2,3, \tag{4.1.3}$$

with the Fourier coefficients

$$f^{(i)\wedge_R}(n,m) = (f^{(i)}, y_{n,m}^{(i),R})_{l^2(\Omega_R)}, \quad i = 1,2,3.$$

Note that (4.1.3) is understood in $\|\cdot\|_{l^2(\Omega_R)}$-sense.
In the same way (as defined in (4.1.2)) we can transfer the system $\{u_{n,m}^{(i)}\}_{\substack{i=1,2,3;\, n\in\mathbb{N}_{0_i};\\ m=1,\ldots,2n+1}}$ to the system $u_{n,m}^{(i),R}(x)_{\substack{i=1,2,3;\, n\in\mathbb{N}_{0_i};\\ m=1,\ldots,2n+1}}$ of vector spherical harmonics for an arbitrary sphere Ω_R by

$$u_{n,m}^{(i),R}(x) = \frac{1}{R} u_{n,m}^{(i)}\left(\frac{x}{|x|}\right), \quad x \in \Omega_R. \tag{4.1.4}$$

These preparations will help us now to introduce the theory for scalar and vector outer harmonics.

4.2 Outer Harmonics

4.2.1 Scalar Outer Harmonics

As point of departure for our formulation of the Runge approximation property on regular surfaces we introduce now scalar outer harmonics as done in [18], [33] which enable us to practice approximation theory on and outside a sphere.
As usual, Ω_R denotes a sphere around the origin with radius $R > 0$, and Ω_R^{ext} its outer space. The system $\{H_{n,m}^R\}_{\substack{n\in\mathbb{N}_0;\\ m=1,\ldots,2n+1}}$, of scalar outer harmonics $H_{n,m}^R$ of degree n and order m defined by

$$H_{n,m}^R(x) = \left(\frac{R}{|x|}\right)^{n+1} Y_{n,m}^R\left(\frac{x}{|x|}\right), \quad x \in \mathbb{R}^3 \setminus \{0\}, \tag{4.2.1}$$

satisfies the following properties:

- $H_{n,m}^R$ is of class $C^{(\infty)}\mathbb{R}^3 \setminus \{0\}$,
- $H_{n,m}^R$ is harmonic in $\mathbb{R}^3 \setminus \{0\}$, i.e.,

$$\Delta_x H_{n,m}^R(x) = 0 \quad x \in \mathbb{R}^3 \setminus \{0\},$$

- $H_{n,m}^R$ is regular at infinity, i.e.,

$$|H_{n,m}^R(x)| = O(|x|^{-1}), \quad |x| \to \infty,$$

and

$$|\nabla_x H^R_{n,m}(x)| = O(|x|^{-2}), \quad |x| \to \infty,$$

- $H^R_{n,m}|_{\Omega_R} = Y^R_{n,m}$,
- $(H^R_{n,m}, H^R_{l,s})_{\mathcal{L}^2(\Omega_R)} = \delta_{n,l}\delta_{m,s}$.

We denote by $Harm_n(\overline{\Omega_R^{ext}})$ the space of all scalar outer harmonics of degree n, i.e.,

$$Harm_n(\overline{\Omega_R^{ext}}) = span_{m=1,\ldots,2n+1}(H^R_{n,m}|\overline{\Omega_R^{ext}}),$$

and $Harm_{p,\ldots,q}(\overline{\Omega_R^{ext}})$, $0 \leq p \leq q$, denotes the space

$$Harm_{p,\ldots,q}(\overline{\Omega_R^{ext}}) = \bigoplus_{n=p}^{q} Harm_n(\overline{\Omega_R^{ext}}).$$

From the addition theorem of scalar spherical harmonics we can deduce the addition theorem of scalar outer harmonics as follows.

Theorem 4.2.1. *Let $\{H^R_{n,m}\}_{\substack{n\in\mathbb{N}_0;\\ m=1,\ldots,2n+1}}$ be a system of scalar outer harmonics of degree n. Then, for any pair $(x,y) \in \overline{\Omega_R^{ext}} \times \overline{\Omega_R^{ext}}$, the addition theorem reads as follows:*

$$\sum_{m=1}^{2n+1} H^R_{n,m}(x)H^R_{n,m}(y) = \frac{2n+1}{4\pi R^2}\left(\frac{R^2}{|x||y|}\right)^{n+1} P_n\left(\frac{x}{|x|}\cdot\frac{y}{|y|}\right).$$

An important property of scalar outer harmonics (see [8, 18]) is that

$$L^2(\Sigma) = \overline{span_{\substack{n\in\mathbb{N}_0;\\ m=1,\ldots,2n+1}}(H^R_{n,m})|_\Sigma}^{\|\cdot\|_{L^2(\Sigma)}},$$

$$C(\Sigma) = \overline{span_{\substack{n\in\mathbb{N}_0;\\ m=1,\ldots,2n+1}}(H^R_{n,m})|_\Sigma}^{\|\cdot\|_{C(\Sigma)}},$$

where Σ is a regular surface (e.g., the real Earth's surface) and Ω_R is a so called Runge sphere inside Σ, i.e., $R < \sigma = inf_{x\in\Sigma}|x|$.

4.2.2 Vector Outer Harmonics

In this section we extend the theory of scalar outer harmonics to the vectorial case (as, e.g., in [14, 33]). We use the system of vector spherical harmonics $\{u^{(i)}_{n,m}\}_{\substack{i=1,2,3;\, n\in\mathbb{N}_{0_i};\\ m=1,2,\ldots,2n+1}}$ in order to generate the set of vector outer harmonics $\{h_{n,m}^{(i),R}\}_{\substack{i=1,2,3;\, n\in\mathbb{N}_{0_i};\\ m=1,2,\ldots,2n+1}}$ in such a way, that the Laplace equation is fulfilled componentwise. These functions when restricted to a regular surface can then be used to approximate continuous vector fields on regular surfaces which is the matter of Chapter 5.

Remark 4.2.1. *The vector outer harmonics are derived formally by applying the Laplace operator on a harmonic function f given by*

$$f(x) = \sum_{i=1}^{3} \sum_{n=0_i}^{\infty} \sum_{m=1}^{2n+1} f_{n,m}^{(i)\wedge}(r) u_{n,m}^{(i)}(\xi),$$

where the Fourier coefficients depending on r are given by

$$f_{n,m}^{(i)\wedge}(r) = \int_{\Omega} f(r\xi) \cdot u_{n,m}^{(i)}(\xi) d\omega(\xi).$$

Using the fact that f fulfills the Laplace equation componentwise and that the vector fields $u_{n,m}^{(i)}$ are eigenfunctions of the (scalar) Beltrami operator, we derive (see [33]) the differential equations

$$\begin{aligned}
\left(\frac{\partial^2}{\partial r^2} + \frac{2}{r}\frac{\partial}{\partial r} - \frac{1}{r^2}(n+1)(n+2) \right) f_{n,m}^{(1)\wedge}(r) &= 0, \\
\left(\frac{\partial^2}{\partial r^2} + \frac{2}{r}\frac{\partial}{\partial r} - \frac{1}{r^2}n(n-1) \right) f_{n,m}^{(2)\wedge}(r) &= 0, \\
\left(\frac{\partial^2}{\partial r^2} + \frac{2}{r}\frac{\partial}{\partial r} - \frac{1}{r^2}n(n+1) \right) f_{n,m}^{(3)\wedge}(r) &= 0,
\end{aligned}$$

$n \in \mathbb{N}_{0_i}$*; $m = 1, \ldots, 2n+1$. The solutions of these differential equations are given by*

$$\begin{aligned}
f_{n,m}^{(1)\wedge}(r) &= \frac{C_1}{r^{n+2}} + D_1 r^{n+1}, \\
f_{n,m}^{(2)\wedge}(r) &= \frac{C_2}{r^{n}} + D_2 r^{n-1}, \\
f_{n,m}^{(3)\wedge}(r) &= \frac{C_3}{r^{n+1}} + D_3 r^{n},
\end{aligned}$$

where C_i, $D_i \in \mathbb{R}$, $i = 1, 2, 3$, are arbitrarily chosen constants. Since we are only interested in the outer space of Ω_R we choose $D_i = 0$ for $i = 1, 2, 3$, in order to fulfill the property of regularity at infinity of the harmonic solutions.

The freedom in choosing the constants implies many possibilities to introduce vector outer harmonics. We will work with vector outer harmonics as presented in [33].
The vector outer harmonics $h_{n,m}^{(i),R}$ of degree n and kind i are defined for $x \in \overline{\Omega_R^{ext}}$ and $n \in \mathbb{N}_{0_i}$; $m = 1, \ldots, 2n+1$, by

$$h_{n,m}^{(1),R}(x) = \left(\frac{R}{|x|}\right)^{n+2} u_{n,m}^{(1),R}\left(\frac{x}{|x|}\right), \tag{4.2.2}$$

$$h_{n,m}^{(2),R}(x) = \left(\frac{R}{|x|}\right)^{n} u_{n,m}^{(2),R}\left(\frac{x}{|x|}\right), \tag{4.2.3}$$

$$h_{n,m}^{(3),R}(x) = \left(\frac{R}{|x|}\right)^{n+1} u_{n,m}^{(3),R}\left(\frac{x}{|x|}\right). \tag{4.2.4}$$

We can give a common representation by setting

$$l_i = \begin{cases} 4, & i = 1, \\ i, & i = 2, 3, \end{cases}$$

i.e., $l_i = i + 3(1 - 0_i)$ with $n \in \mathbb{N}_{0_i}$; $m = 1, \ldots, 2n+1$; $i = 1, 2, 3$,

$$h_{n,m}^{(i),R}(x) = \frac{1}{R}\left(\frac{R}{|x|}\right)^{n-2+l_i} u_{n,m}^{(i)}\left(\frac{x}{|x|}\right). \tag{4.2.5}$$

The following properties are satisfied [33]:

- $h_{n,m}^{(i),R}$ is of class $c^{(\infty)}(\Omega_R^{ext})$,
- $\Delta_x h_{n,m}^{(i),R}(x) = 0$ for $x \in \Omega_R^{ext}$, i.e., every component function $h_{n,m}^{(i),R} \cdot \varepsilon^k$, $k = 1, 2, 3$ satisfies the Laplace equation,
- $h_{n,m}^{(i),R}|_{\Omega_R} = \frac{1}{R} u_{n,m}^{(i)}$,
- $(h_{n,m}^{(i),R}, h_{l,s}^{(j),R})_{l^2(\Omega_R)} = \delta_{ij}\delta_{nl}\delta_{ms}$,
- $|h_{n,m}^{(i),R}(x)| = O(|x|^{-1}), \quad |x| \to \infty, \quad x \in \Omega_R^{ext}$.

Similar to the scalar outer harmonics we define the space for vector outer harmonics by

$$harm_n^{(i)}(\overline{\Omega_R^{ext}}) = span\{h_{n,m}^{(i),R} \,|\, m = 1, \ldots, 2n+1\}$$

and

$$\begin{aligned} harm_0(\overline{\Omega_R^{ext}}) &= harm_0^{(1)}(\overline{\Omega_R^{ext}}), \\ harm_n(\overline{\Omega_R^{ext}}) &= \bigoplus_{i=1}^{3} harm_n^{(i)}(\overline{\Omega_R^{ext}}), \end{aligned}$$

where Ω_R^{ext} again denotes the outer space of a sphere Ω_R. The representation for bandlimited spaces $harm_{p,\ldots,q}^{(i)}(\overline{\Omega_R^{ext}})$, $0_i \le p \le q$, is naturally given by

$$harm_{p,\ldots,q}^{(i)}(\overline{\Omega_R^{ext}}) = \bigoplus_{n=p}^{q} harm_n^{(i)}(\overline{\Omega_R^{ext}}),$$

whereas we have

$$harm(\overline{\Omega_R^{ext}}) = \overline{span_{\substack{i=1,2,3;\, n\in\mathbb{N}_{0_i};\\ m=1,2,\ldots,2n+1}} h_{n,m}^{(i),R}}^{\|\cdot\|_{c(\overline{\Omega_R^{ext}})}} \tag{4.2.6}$$

and

$$harm^{(i)}(\overline{\Omega_R^{ext}}) = \overline{span_{\substack{n\in\mathbb{N}_{0_i};\\ m=1,2,\ldots,2n+1}} h_{n,m}^{(i),R}}^{\|\cdot\|_{c(\overline{\Omega_R^{ext}})}}. \tag{4.2.7}$$

Using the Legendre tensors given by (3.3.24) we are able to deduce the addition theorem for vector outer harmonics from the addition theorem for vector spherical harmonics $u_{n,m}^{(i)}$.

Theorem 4.2.2. *Let* $\{h_{n,m}^{(i),R}\}_{\substack{i=1,2,3;n\in\mathbb{N}_{0_i};\\ m=1,\ldots,2n+1}}$, *be a system of vector outer harmonics of degree n and kind i as defined by (4.2.5). Then for any pair* $(x,y) \in \overline{\Omega_R^{ext}} \times \overline{\Omega_R^{ext}}$ *the addition theorem for vector outer harmonics reads as follows:*

$$\sum_{m=1}^{2n+1} h_{n,m}^{(i),R}(x) \otimes h_{n,m}^{(j),R}(y) = \frac{1}{R^2}\left(\frac{R}{|x|}\right)^{n-2+l_i}\left(\frac{R}{|y|}\right)^{n-2+l_j}\frac{2n+1}{4\pi}\tilde{\mathbf{p}}_n^{(i,j)}(\xi,\eta),$$

where $l_i = i + 3(1-0_i)$, $i = 1,2,3$, *and* $l_j = j + 3(1-0_i)$, $j = 1,2,3$ *and the* $\tilde{\mathbf{p}}_n^{(i,j)}$ *are given by (3.3.24).*

4.2.3 Closure of Vector Outer Harmonics

Finally we want to discuss the closure properties of vector outer harmonics on regular surfaces. Thus we will show that linear independence is one important property of vector outer harmonics.

Lemma 4.2.3. *Let Σ be a regular surface and $\{h_{n,m}^{(i),R}\}_{\substack{i=1,2,3;n\in\mathbb{N}_{0_i};\\ m=1,\ldots,2n+1}}$ be a system of vector outer harmonics of degree n and kind i as given in (4.2.2)-(4.2.4). Then the system $\{h_{n,m}^{(i),R}|_\Sigma\}_{\substack{i=1,2,3;n\in\mathbb{N}_{0_i};\\ m=1,\ldots,2n+1}}$ is linearly independent.*

The proof can be found in [33] based on results in [8].
The following lemma (see, e.g., [33]) is an auxiliary tool to prove the theorem of completeness for vector outer harmonics. It follows from the corresponding results of scalar outer harmonics.

Lemma 4.2.4. *Let $\{H_{n,m}^R\}_{n\in\mathbb{N}_0;\, m=1,\ldots,2n+1}$ be a system of scalar outer harmonics as defined by (4.2.1). Then*

$$\begin{aligned} \overline{span_{\substack{i=1,2,3;n\in\mathbb{N}_0;\\ m=1,\ldots,2n+1}}\{H_{n,m}^R\varepsilon^i|_\Sigma\}}^{\|\cdot\|_{l^2(\Sigma)}} &= l^2(\Sigma), \\ \overline{span_{\substack{i=1,2,3;n\in\mathbb{N}_0;\\ m=1,\ldots,2n+1}}\{H_{n,m}^R\varepsilon^i|_\Sigma\}}^{\|\cdot\|_{c(\Sigma)}} &= c(\Sigma). \end{aligned}$$

In the following theorem a completeness result for the vector outer harmonics is given (see [33]).

Theorem 4.2.5. *Let $\{h_{n,m}^{(i),R}\}_{\substack{i=1,2,3;n=0_i,1,\ldots;\\ m=1,\ldots,2n+1}}$ be a system of vector outer harmonics as defined by (4.2.5). Then the following statements hold true:*

$$\overline{span_{\substack{i=1,2,3;n\in\mathbb{N}_{0_i};\\ m=1,\ldots,2n+1}}\{h_{n,m}^{(i),R}|_\Sigma\}}^{\|\cdot\|_{l^2(\Sigma)}} = l^2(\Sigma), \tag{4.2.8}$$

$$\overline{span_{\substack{i=1,2,3;n\in\mathbb{N}_{0_i};\\ m=1,\ldots,2n+1}}\{h_{n,m}^{(i),R}|_\Sigma\}}^{\|\cdot\|_{c(\Sigma)}} = c(\Sigma). \tag{4.2.9}$$

Proof. Let $\varepsilon^k H_{n,m}^R$ be a homogeneous harmonic vector polynomial for any fixed value of $k=1,2,3$; $n\in\mathbb{N}_0$; $m=0,\ldots 2n+1$. Then this homogeneous harmonic vector polynomial can be written as a linear combination of vector spherical harmonics, given by Lemma 3.3.12. Let us first represent $\varepsilon^k H_{n,m}^R$ restricted to the unit sphere Ω_R with radius $R=1$, i.e.,

$$\varepsilon^k H_{n,m}^R(x)|_\Omega = u_{n-1}^{(1)}\left(\frac{x}{|x|}\right) + u_{n+1}^{(2)}\left(\frac{x}{|x|}\right) + u_n^{(3)}\left(\frac{x}{|x|}\right), \quad x\in\Omega_R,\ R=1.$$

By suspending the restriction on the unit sphere thus allowing $R\geq 1$ we get the following representation:

$$\begin{aligned}
&\varepsilon^k H^R_{n,m}(x)\\
= &\left(\frac{R}{|x|}\right)^{n+1}\left(u^{(1),R}_{n-1}\left(\frac{x}{|x|}\right)+u^{(2),R}_{n+1}\left(\frac{x}{|x|}\right)+u^{(3),R}_{n}\left(\frac{x}{|x|}\right)\right)\\
= &\sum_{m=1}^{2(n-1)+1} a^{(1)}_{n-1,m}h^{(1),R}_{n-1,m}(x)+\sum_{m=1}^{2(n+1)+1} a^{(2)}_{n+1,m}h^{(2),R}_{n+1,m}(x)+\sum_{m=1}^{2n+1} a^{(3)}_{n,m}h^{(3),R}_{n,m}(x).
\end{aligned}$$

Therefore,

$$\begin{aligned}
&span_{\substack{k=1,2,3;\ n\in\mathbb{N}_0;\\ m=0,\ldots 2n+1}}\{\varepsilon^k H^R_{n,m}|_\Sigma\}\\
\subset\ &\bigoplus_{m=1}^{2(n-1)+1}\{h^{(1),R}_{n-1,m}|_\Sigma\}\ \oplus\ \bigoplus_{m=1}^{2(n+1)+1}\{h^{(2),R}_{n+1,m}|_\Sigma\}\ \oplus\ \bigoplus_{m=1}^{2(n)+1}\{h^{(3),R}_{n,m}|_\Sigma\}.
\end{aligned}$$

Lemma 4.2.4 completes the proof. □

Thus the system $\{h^{(i),R}_{n,m}\}_{\substack{i=1,2,3;n\in\mathbb{N}_{0_i};\\ m=1,\ldots,2n+1}}$ of vector outer harmonics is orthogonal when restricted to a sphere Ω_R linearly independent and complete when restricted to a regular surface Σ. From functional analysis (see e.g. [3]) we know that in a Hilbert space such as $l^2(\Sigma)$ the properties of completeness and closure are equivalent. This leads to the following corollary.

Corollary 4.2.6. *Under the assumption of Theorem 4.2.5 the following statement is valid. The system of vector outer harmonics* $\{h^{(i),R}_{n,m}|_\Sigma\}_{\substack{i=1,2,3;n=0_i,1,\ldots;\\ m=1,\ldots,2n+1}}$ *is closed in* $l^2(\Sigma)$, *i.e., for given* $f\in l^2(\Sigma)$ *and arbitrary* $\varepsilon>0$ *there exists a linear combination*

$$h_m=\sum_{i=1}^{3}\sum_{n=0_i}^{N}\sum_{m=1}^{2n+1} a^{(i)}_{nm}h^{(i),R}_{n,m}|_\Sigma,$$

such that

$$\|f-h_m\|_{l^2(\Sigma)}\le\varepsilon,$$

where $a^{(i)}_{nm}$ *are appropriate coefficients.*

4.3 Exact Computation of Homogeneous Harmonic Polynomials

Two algorithms exclusively using on integer operations are explained for establishing linearly independent systems of homogeneous harmonic polynomials. The first algorithm is based on the solution of an underdetermined system of linear equations, whereas the second algorithm uses a recursion relation for two-dimensional homogeneous polynomials.

4.3.1 Exact Computation Via Underdetermined Linear Systems

Our purpose now is to explain how a maximal linearly independent and orthonormal system of homogeneous harmonic polynomials of degree n can be generated exactly (our considerations literally follow [19]). The concept is based on the observation that any linearly independent system $\{H_{n,j}\}_{j=1,\ldots,2n+1}$ of homogeneous harmonic polynomials of degree n

$$\begin{aligned} H_{n,1}(x) &= \sum_{[\alpha]=n} C^1_\alpha x^\alpha \\ \vdots\ \vdots & \qquad \vdots\ \vdots \\ H_{n,2n+1}(x) &= \sum_{[\alpha]=n} C^{2n+1}_\alpha x^\alpha \end{aligned} \tag{4.3.1}$$

can be determined by *exact computation* of the coefficients C^j_α, i.e., entirely by integer operations for $j = 1, \ldots, 2n+1$. (Note that we briefly write C^j_α instead of $C^{n,j}_\alpha$ when confusion is not likely to arise). In other words, we want to show that the coefficients C^j_α, $j = 1, ..., 2n+1$, in (4.3.1) can be expressed as integers.

Generation of Linearly Independent Systems

Let H_n be a homogeneous polynomial of the form $H_n = \Sigma_{[\alpha]=n} C_\alpha x^\alpha$, $x \in \mathbb{R}^3$, $n \geq 2$. Assuming that H_n is harmonic, i.e., $\Delta_x H_n(x) = 0$, $x \in \mathbb{R}^3$, we obtain

$$\Delta_x H_n(x) = \Delta_x \sum_{[\alpha]=n} C_\alpha x^\alpha = \sum_{[\alpha]=n} C_\alpha \Delta_x(x^\alpha) = 0. \tag{4.3.2}$$

Thus it follows that

$$\begin{aligned} \sum_{\alpha_1+\alpha_2+\alpha_3=n} & C_\alpha\big(\alpha_1(\alpha_1-1)x_1^{\alpha_1-2}x_2^{\alpha_2}x_3^{\alpha_3} + \alpha_2(\alpha_2-1)x_2^{\alpha_2-2}x_1^{\alpha_1}x_3^{\alpha_3} \\ & + \alpha_3(\alpha_3-1)x_3^{\alpha_3-2}x_1^{\alpha_1}x_2^{\alpha_2}\big) = 0. \end{aligned} \tag{4.3.3}$$

We discuss the terms

$$\alpha_1(\alpha_1-1)x_1^{\alpha_1-2}x_2^{\alpha_2}x_3^{\alpha_3}, \ \alpha_1+\alpha_2+\alpha_3 = n,$$

$$\alpha_2(\alpha_2-1)x_1^{\alpha_1}x_2^{\alpha_2-2}x_3^{\alpha_3},\ \alpha_1+\alpha_2+\alpha_3=n, \tag{4.3.4}$$
$$\alpha_3(\alpha_3-1)x_1^{\alpha_1}x_2^{\alpha_2}x_3^{\alpha_3-2},\ \alpha_1+\alpha_2+\alpha_3=n$$

in more detail. Every term in (4.3.4) with index $\alpha = (\alpha_1, \alpha_2, \alpha_3)^T$ satisfying $[\alpha] = \alpha_1 + \alpha_2 + \alpha_3 = n$ is a homogeneous polynomial of degree $n-2$. Hence, the left hand side of (4.3.3) is a homogeneous polynomial of degree $n-2$. Therefore ΔH_n can be represented in the form

$$\Delta_x H_n(x) = \sum_{[\beta]=n-2} D_\beta x^\beta. \tag{4.3.5}$$

The coefficients D_β are given by

$$D_\beta = \sum_{[\alpha]=n} C_\alpha m_{\beta\alpha}, \tag{4.3.6}$$

where $m_{\beta\alpha}$ is given by

$$m_{\beta\alpha} = \begin{cases} \alpha_1(\alpha_1-1), & \beta-\alpha = (-2,0,0)^T \\ \alpha_2(\alpha_2-1), & \beta-\alpha = (0,-2,0)^T \\ \alpha_3(\alpha_3-1), & \beta-\alpha = (0,0,-2)^T \\ 0, & \textit{otherwise.} \end{cases} \tag{4.3.7}$$

H_n is assumed to be harmonic, i.e., $\Delta_x H_n(x) = 0$ for all $x \in \mathbb{R}^3$. But this means that all numbers D_β are equal to 0. Therefore it follows that

$$\sum_{[\alpha]=n} C_\alpha m_{\beta\alpha} = 0 \tag{4.3.8}$$

for all β with $[\beta] = n-2$. Now, (4.3.8) is a linear system of $\binom{n}{2}$ equations in the $\binom{n+2}{2}$ unknowns C_α, $[\alpha] = n$.

The matrix $\mathbf{m} = (m_{\beta\alpha})$ has $\binom{n}{2}$ rows and $\binom{n+2}{2}$ columns; $\mathbf{m}$ can be partitioned as follows:

$$\mathbf{m} = (\underbrace{\mathbf{l}}_{\binom{n}{2}} \ \vdots \ \underbrace{\mathbf{r}}_{\binom{n+2}{2}-\binom{n}{2}=2n+1})\} \binom{n}{2}, \tag{4.3.9}$$

where $\mathbf{l} = (l_{\beta\delta})$ is a $\binom{n}{2}$ by $\binom{n}{2}$ matrix and $\mathbf{r} = (r_{\beta\delta})$ is a $\binom{n}{2}$ by $\binom{n+2}{2} - \binom{n}{2}$ matrix.

For the set of multiindices of degree n we introduce a binary relation (lexicographical order) between elements

$$\alpha' = (\alpha_1', \alpha_2', \alpha_3')^T,\ \alpha'' = (\alpha_1'', \alpha_2'', \alpha_3'')^T \tag{4.3.10}$$

designated by " $>$ " and defined as follows:

$$\alpha' > \alpha'' \tag{4.3.11}$$

if and only if one of the following relations is satisfied

$$\alpha_1' > \alpha_1'' \tag{4.3.12}$$

or

$$\alpha_1' = \alpha_1'',\ \alpha_2' > \alpha_2'' \tag{4.3.13}$$

or

$$\alpha_1' = \alpha_1'',\ \alpha_2' = \alpha_2'',\ \alpha_3' > \alpha_3''. \tag{4.3.14}$$

The binary relation " $>$ " implies an ordering for the multiindices α, $[\alpha] = n$, according to the mapping

$$\begin{array}{ll}
(n,0,0) & \to 1 \,\}\, 1 \\
\left.\begin{array}{l}(n-1,1,0) \to 2 \\ (n-1,0,1) \to 3\end{array}\right\} & 2 \\
\left.\begin{array}{l}(n-2,2,0) \to 4 \\ (n-2,1,1) \to 5 \\ (n-2,0,2) \to 6\end{array}\right\} & 3 \\
\vdots\ \vdots\ \vdots & \\
\left.\begin{array}{l}(0,n,0) \to \binom{n+2}{2} - n \\ \vdots\ \vdots\ \vdots \\ (0,0,n) \to \binom{n+2}{2}\end{array}\right\} & n+1\,.
\end{array}$$

In the same way, the set of multiindices β, $[\beta] = n-2$, may be ordered by increasing integers $i, 1 \leq i \leq \binom{n}{2}$. Hence, in canonical manner, each pair (β, α) with $[\beta] = n-2$, $[\alpha] = n$, corresponds uniquely to a pair $(i,j), 1 \leq i \leq \binom{n}{2}, 1 \leq j \leq \binom{n+2}{2}$. In this notation the matrix

$$\mathbf{m} \;=\; (m_{\beta\alpha}),\ [\beta] \;=\; n-2,\ [\alpha] \;=\; n, \tag{4.3.15}$$

can be rewritten in the ordered form

$$\mathbf{m} = (m_{ij}),\ 1 \leq i \leq \binom{n}{2},\ 1 \leq j \leq \binom{n+2}{2}. \tag{4.3.16}$$

Analogously

$$\mathbf{l} = l_{\beta\gamma}, \ [\beta] = n-2, \ [\gamma] = n-2 \tag{4.3.17}$$

becomes

$$\mathbf{l} = (l_{ij}), \ 1 \le i \le \binom{n}{2}, \ 1 \le j \le \binom{n}{2}. \tag{4.3.18}$$

¿From (4.3.7) it can be deduced that

$$\begin{aligned} l_{ij} &= 0 \quad for \quad i > j, \ i = 2, ..., \tbinom{n}{2}, \\ l_{ij} &\neq 0 \quad for \quad i = j, \ i = 1, ..., \tbinom{n}{2}. \end{aligned} \tag{4.3.19}$$

But this shows that $\mathbf{l}$ is non-singular, hence, the matrix $\mathbf{m}$ is of maximal rank: $\binom{n}{2}$. Therefore we are able to find $\binom{n+2}{2} - \binom{n}{2}$, i.e., $2n+1$ linearly independent solution vectors (A^1_α) , ... , (A^{2n+1}_α), $[\alpha] = n$, of the homogeneous linear system (4.3.8). According to standard arguments of linear algebra the $\binom{n+2}{2}$ by $2n+1$ matrix $\mathbf{a}$ consisting of the vectors $(A^1_\alpha), ..., (A^{2n+1}_\alpha)$

$$\mathbf{a} = \underbrace{((A^1_\alpha), ..., (A^{2n+1}_\alpha))}_{2n+1} \ \Big\} \ \tbinom{n+2}{2} \tag{4.3.20}$$

may be partitioned in the following form

$$\mathbf{a} = \begin{pmatrix} \mathbf{u} \\ -\mathbf{i} \end{pmatrix}, \tag{4.3.21}$$

where $\mathbf{i}$ is the $(2n+1)$ by $(2n+1)$ unit matrix, and $\mathbf{u}$ is a $\binom{n+2}{2}$ by $(2n+1)$ by $(2n+1)$ matrix. Then the linear system $\mathbf{m} \ \mathbf{a} = \mathbf{0}$ can be written as follows $\mathbf{l} \ \mathbf{u} = \mathbf{r}$. Since $\mathbf{l}$ is a $(2n+1)$ by $(2n+1)$ upper triangular matrix, the unknown matrix $\mathbf{u}$ can be computed by $(2n+1)$-times backward substitution.

The elements of the matrix $\mathbf{m} = (m_{\beta\alpha})$ are all integers. Therefore, any solution of the linear system (4.3.8) is a column vector of rational components. Hence, there exists a matrix

$$\mathbf{c} = ((C^1_\alpha), ..., (C^{2n+1}_\alpha)), \ [\alpha] = n, \tag{4.3.22}$$

the elements of which are all integers (observe that if (C_α), $[\alpha] = n$, is a solution of (4.3.8), then $k \ (C_\alpha)$, $[\alpha] = n$, k integer, is a solution, too).

In other words, the solution process can be performed strictly in the modulus of integers. Exact computation (without rounding errors) is possible in integer mode by use of integer operations (addition, subtraction, multiplication of integers). When the matrix $\mathbf{c}$ has been calculated, the homogeneous harmonic polynomials $H_{n,j}$ given by (4.3.1) form a (maximal) linearly independent system, i.e., a basis in $Harm_n$.

Finally, it should be emphasized that exact computation, i.e., addition, subtraction, multiplication in integer mode must be performed strictly in the available range of the integer constants. Helpful is an arithmetic for arbitrarily long integers whose implementation on a computer system operates with lists so that there is no restriction on the size of the integers worked with (this is a standard feature of computer algebra packages). Let us demonstrate the technique of calculating the matrix $\mathbf{c}$ for two examples:

We choose the degree $n = 3$. Then an elementary calculation yields

$$\binom{n+2}{2} = 10, \quad \binom{n}{2} = 3, \tag{4.3.23}$$

hence,

$$\binom{n+2}{2} - \binom{n}{2} = 7. \tag{4.3.24}$$

Every homogeneous harmonic polynomial $H_3 \in \mathrm{Hom}_3$ may be represented in the form:

$$\begin{aligned} H_3(x) &= C_{300}\, x_1^3 + C_{210}\, x_1^2 x_2 + C_{201}\, x_1^2 x_3 \\ &+ C_{120}\, x_1 x_2^2 + C_{111}\, x_1 x_2 x_3 + C_{102}\, x_1 x_3^2 \\ &+ C_{030}\, x_2^3 + C_{021}\, x_2^2 x_3 + C_{012}\, x_2 x_3^2 \\ &+ C_{003}\, x_3^3 \\ &\qquad (x = (x_1, x_2, x_3)^T). \end{aligned} \tag{4.3.25}$$

H_3 has to fulfill the differential equation $\Delta_x H_3(x) = 0,\ x \in \mathbb{R}^3$, i.e.,

$$\begin{aligned} 6\, C_{300} x_1 &+ 2\, C_{210} x_2 + 2\, C_{201} x_3 \\ +2\, C_{120} x_1 &+ 6\, C_{030} x_2 + 2\, C_{021} x_3 \\ +2\, C_{102} x_1 &+ 2\, C_{012} x_2 + 6\, C_{003} x_3 = 0. \end{aligned} \tag{4.3.26}$$

Since $\Delta_x H_3(x) = 0$ identically for all $x \in \mathbb{R}^3$ we get $\binom{n}{2} = 3$ equations for the coefficients

$$\begin{aligned} 6C_{300} &+ 2C_{120} + 2C_{102} = 0, \\ 2C_{210} &+ 6C_{030} + 2C_{012} = 0, \\ 2C_{201} &+ 2C_{021} + 6C_{003} = 0. \end{aligned}$$

Using the introduced order for the coefficients C_α, $[\alpha] = 3$, the equation $\mathbf{m}\,\mathbf{c} = 0$ reads in matrix notation

$$\begin{pmatrix} 6 & 0 & 0 & \vdots & 2 & 0 & 2 & 0 & 0 & 0 & 0 \\ 0 & 2 & 0 & \vdots & 0 & 0 & 0 & 6 & 0 & 2 & 0 \\ 0 & 0 & 2 & \vdots & 0 & 0 & 0 & 0 & 2 & 0 & 6 \end{pmatrix} \begin{pmatrix} C_{300} \\ C_{210} \\ C_{201} \\ \dots \\ C_{120} \\ C_{111} \\ C_{102} \\ C_{030} \\ C_{021} \\ C_{012} \\ C_{003} \end{pmatrix} = \begin{pmatrix} 0 \\ 0 \\ 0 \\ \dots \\ 0 \\ 0 \\ 0 \\ 0 \\ 0 \\ 0 \\ 0 \end{pmatrix}, \tag{4.3.27}$$

where we have marked the partitioning of the matrix $\mathbf{m}$ and the vector (C_α) by dashed lines. If we choose

$$C_{120} = -1, C_{111} = \dots = C_{003} \;=\; 0 \tag{4.3.28}$$

the linear system is uniquely solved by the vector

$$(\frac{1}{3}, 0, 0 \,\vdots\, -1, 0, 0, 0, 0, 0, 0)^T. \tag{4.3.29}$$

Multiplying this vector by 3 all components become integers

$$(C_\alpha^1) = (1, 0, 0 \,\vdots\, -3, 0, 0, 0, 0, 0, 0)^T. \tag{4.3.30}$$

In the same way we generate a set of 7 linearly independent solutions of the above system the components of which are all integers, viz.

$$\begin{aligned}
(C_\alpha^2) &= (0, 0, 0 \,\vdots\, 0, -1, 0, 0, 0, 0, 0)^T, \\
(C_\alpha^3) &= (1, 0, 0 \,\vdots\, 0, 0, -3, 0, 0, 0, 0)^T, \\
(C_\alpha^4) &= (0, 3, 0 \,\vdots\, 0, 0, 0, -1, 0, 0, 0)^T, \\
(C_\alpha^5) &= (0, 0, 1 \,\vdots\, 0, 0, 0, 0, -1, 0, 0)^T, \\
(C_\alpha^6) &= (0, 1, 0 \,\vdots\, 0, 0, 0, 0, 0, -1, 0)^T, \\
(C_\alpha^7) &= (0, 0, 3 \,\vdots\, 0, 0, 0, 0, 0, 0, -1)^T.
\end{aligned}$$

Thus a linearly independent system $\{H_{3,j}\}_{j=1,\dots,7}$ of homogeneous harmonic polynomials of degree 3 is found by the following functions:

$$\begin{aligned}
H_{3,1}(x) &= 1 \cdot x_1^3 - 3 \cdot x_1 x_2^2, \\
H_{3,2}(x) &= -1 \cdot x_1 x_2 x_3, \\
H_{3,3}(x) &= 1 \cdot x_1^3 - 3 \cdot x_1 x_3^2, \\
H_{3,4}(x) &= 3 \cdot x_1^2 x_2 - 1 \cdot x_2^3, \\
H_{3,5}(x) &= 1 \cdot x_1^2 x_3 - 1 \cdot x_2^2, \\
H_{3,6}(x) &= 1 \cdot x_1^2 x_2 - 1 \cdot x_2 x_3^2, \\
H_{3,7}(x) &= 3 \cdot x_1^2 x_3 - 1 \cdot x_3^3.
\end{aligned}$$

Generation of Orthonormal Systems

The linearly independent systems of homogeneous harmonic polynomials as developed here turn out to be of particular significance when outer harmonics have to be orthonormalized with respect to a (geoscientifically relevant) regular surface Σ, i.e., when a linearly independent and not an $\mathcal{L}^2(\Omega)$-orthonormal system is required for the orthonormalization process on Σ. Nevertheless it is worth mentioning that, corresponding to the linearly independent system $\{H_{n,m}\}_{m=1,\dots,2n+1}$ of homogeneous harmonic polynomials of degree n, an orthogonal system, in $\{H^*_{n,m}\}_{m=1,\dots,2n+1}$ with respect to both the topology of $\mathrm{H}om_n$ and $\mathcal{L}^2(\Omega)$ can be constructed only by integer operations (according to the well-known Gram-Schmidt process). To this end the functions $H^*_{n,m}$ are computed recursively. We start from

$$H^*_{n,1} = H_{n,1} \ . \tag{4.3.31}$$

Then we set

$$H^*_{n,2} = a^n_{2,1} H^*_{n,1} + H_{n+1} \ . \tag{4.3.32}$$

The coefficient $a^n_{2,1}$ has to be chosen such that $H^*_{n,2}$ is orthogonal to $H^*_{n,1}$:

$$(H^*_{n,2}, H^*_{n,1})_{Hom_n} = 0 \ . \tag{4.3.33}$$

It turns out that

$$a^n_{2,1} = -\frac{(H_{n,2}, H^*_{n,1})_{Hom_n}}{(H^*_{n,1}, H^*_{n,1})_{Hom_n}} \ . \tag{4.3.34}$$

It should be noted that numerator and denominator may be determined exactly. Now, let

$$H^*_{n,3} = a^n_{3,1} H^*_{n,1} + a^n_{3,2} H^*_{n,2} + H_{n,3} \; . \tag{4.3.35}$$

The requirements

$$\begin{aligned} (H^*_{n,3}, H^*_{n,1})_{Hom_n} &= 0, \\ (H^*_{n,3}, H^*_{n,2})_{Hom_n} &= 0 \end{aligned}$$

lead to

$$a^n_{3,1} = -\frac{(H_{n,3}, H^*_{n,1})_{Hom_n}}{(H^*_{n,1}, H^*_{n,1})_{Hom_n}} \; , \tag{4.3.36}$$

$$a^n_{3,2} = -\frac{(H_{n,3}, H^*_{n,2})_{Hom_n}}{(H^*_{n,2}, H^*_{n,2})_{Hom_n}} \; . \tag{4.3.37}$$

Again, the coefficients can be deduced by integer operations. Analogously we get, in general,

$$H^*_{n,k} = a^n_{k,1} H^*_{n,1} + \ldots + a^n_{k,k-1} H^*_{n,k-1} + H_{n,k}, \quad k = 2, \ldots, 2n+1, \tag{4.3.38}$$

$$H^*_{n,1} = H_{n,1} \; , \tag{4.3.39}$$

where the coefficients

$$a^n_{k,s} = -\frac{(H_{n,k}, H^*_{n,s})_{Hom_n}}{(H^*_{n,s}, H^*_{n,s})_{Hom_n}} \; . \tag{4.3.40}$$

are computable exactly by integer operations, i.e., $a^n_{k,s}$ is known exactly as a fraction of integers.

According to this well–known orthogonalization scheme, each function $H^*_{n,j}$ is a linear combination of the functions $H_{n,1}, \ldots, H_{n,2n+1}$. The coefficients of this linear combination can be obtained exactly as rational numbers, too. Thus there exists a vector (B^m_α) such that

$$H^*_{n,m}(x) = \sum_{[\alpha]=n} B^m_\alpha x^\alpha \; , \quad m = 1, \ldots, 2n+1 \; . \tag{4.3.41}$$

The vectors (B^m_α), $m = 1, \ldots, 2n+1$, form a matrix $\mathbf{b}$ whose elements consist of fractions of integers (provided that all numbers in the course of computation have been calculated in such a way that numerator and denominator are known as integers).

Lemma 4.3.1. *There exists a sequence of homogeneous harmonic polynomials* $\{H^*_{n,m}\}_{m=1,\ldots,2n+1}$ *of degree* n *with*

$$(H^*_{n,m}, H^*_{n,l})_{Hom_n} = 0, \qquad m \neq l \ ,$$

viz.

$$\begin{aligned} H^*_{n,1} &= H_{n,1} \\ H^*_{n,k} &= a^n_{k,1} H^*_{n,1} + \ldots + a^n_{k,k-1} H^*_{n,k-1} + H_{n,k}, \quad k = 2, \ldots, 2n+1, \end{aligned}$$

where all coefficients $a^n_{k,s}$ *are computable by integer operations.*

Remark 4.3.1. *Provided that the expression* $\sqrt{(H^*_{n,m}, H^*_{n,m})_{Hom_n}}$ *has been stored as the radicant of an integer a* Hom_n*-orthonormal system of homogeneous harmonic polynomials of degree* n *can be calculated exactly, i.e., by integer operations. The exactness is of basic importance. In fact, the method is constructed so as to avoid computational errors.*

Lemma 4.3.2. *The system*

$$\sqrt{(H^*_{n,1}, H^*_{n,1})^{-1}_{Hom_n}} H^*_{n,1}, \ldots, \ \sqrt{(H^*_{n,2n+1}, H^*_{n,2n+1})^{-1}_{Hom_n}} H^*_{n,2n+1}$$

is an orthonormal system of homogeneous harmonic polynomials of degree n *with respect to* $(\cdot,\cdot)_{Hom_n}$*, while*

$$\sqrt{\mu_n (H^*_{n,1}, H^*_{n,1})^{-1}_{Hom_n}} H^*_{n,1}, \ldots, \ \sqrt{\mu_n (H^*_{n,2n+1}, H^*_{n,2,+1})^{-1}_{Hom_n}} H^*_{n,2n+1}$$

is an orthonormal system of homogeneous harmonic polynomials of degree n *with respect to* $(\cdot,\cdot)_{\mathcal{L}^2(\Omega)}$*. The values* $(H^*_{n,m}, H^*_{n,m})_{Hom_n}$ *can be determined entirely by integer operations.*

We only deal with the degree $n = 3$ (for a table of higher degrees see [9]). According to our orthonormalization process due to Gram-Schmidt we are able to deduce from the maximal system of linearly independent homogeneous harmonic polynomials $\{H_{3,m}\}_{m=1,\ldots,7}$ an orthogonal system $\{H^*_{3,m}\}_{m=1,\ldots,7}$. The resulting functions are listed below:

$$\begin{aligned} H^*_{3,1}(x) &= x_1^3 - 3x_1x_2^2, \\ H^*_{3,2}(x) &= x_1x_2x_3, \\ H^*_{3,3}(x) &= x_1^3 + x_1x_2^2 - 4x_1x_3^2, \\ H^*_{3,4}(x) &= 3x_1^2x_2 - x_2^3 - x_2^3, \\ H^*_{3,5}(x) &= x_1^2x_3 - x_2^2x_3, \\ H^*_{3,6}(x) &= x_1^2x_2 + x_2^3 - 4x_2x_3^2, \\ H^*_{3,7}(x) &= 3x_1^2x_3 + 3x_2^2x_3 - 2x_3^3. \end{aligned}$$

That means, all components $B_\alpha^j \neq 0$ are decomposed into an integer times a product of the prime numbers 2, 3. An easy calculation gives

$$\begin{array}{rclcl} (H^*_{3,1}, H^*_{3,1})_{\mathrm{Hom}_3} & = & 24 & = & 1 \cdot 2^3 \cdot 3^1, \\ (H^*_{3,2}, H^*_{3,2})_{\mathrm{Hom}_3} & = & 1 & = & 1 \cdot 2^0 \cdot 3^0, \\ (H^*_{3,3}, H^*_{3,3})_{\mathrm{Hom}_3} & = & 40 & = & 5 \cdot 2^3 \cdot 3^0, \\ (H^*_{3,4}, H^*_{3,4})_{\mathrm{Hom}_3} & = & 24 & = & 1 \cdot 2^3 \cdot 3^1, \\ (H^*_{3,5}, H^*_{3,5})_{\mathrm{Hom}_3} & = & 4 & = & 1 \cdot 2^2 \cdot 3^0, \\ (H^*_{3,6}, H^*_{3,6})_{\mathrm{Hom}_3} & = & 40 & = & 5 \cdot 2^3 \cdot 3^0, \\ (H^*_{3,7}, H^*_{3,7})_{\mathrm{Hom}_3} & = & 60 & = & 5 \cdot 2^2 \cdot 3^1. \end{array} \tag{4.3.42}$$

Thus, the integers are decomposed into a (positive) integer times a product of prime numbers ≤ 3.

Consequently, the orthonormal system

$$\sqrt{(H^*_{n,m}, H^*_{n,m})^{-1}_{\mathrm{Hom}_3}} H^*_{n,m} \tag{4.3.43}$$

(with respect to $(\cdot,\cdot)_{\mathrm{Hom}_3}$). corresponding to $\{H^*_{n,m}\}_{m=1,\ldots,7}$ may be listed as follows:

$$\begin{aligned} &\sqrt{(H^*_{3,1}, H^*_{3,1})^{-1}_{\mathrm{Hom}_3}}\, H^*_{3,1}(x) \\ &= (1 \cdot 2^0 \cdot 3^0 \cdot x_1^3 x_2^0 x_3^0 - 1 \cdot 2^0 \cdot 3^1 \cdot x_1^1 x_2^2 x_3^0)/\sqrt{1 \cdot 2^3 \cdot 3^1}, \end{aligned}$$

$$\begin{aligned} &\sqrt{(H^*_{3,2}, H^*_{3,2})^{-1}_{\mathrm{Hom}_3}}\, H^*_{3,2}(x) \\ &= (1 \cdot 2^0 \cdot 3^0 \cdot x_1^1 x_2^1 x_3^1)/\sqrt{1 \cdot 2^0 \cdot 3^0}, \end{aligned}$$

$$\begin{aligned} &\sqrt{(H^*_{3,3}, H^*_{3,3})^{-1}_{\mathrm{Hom}_3}}\, H^*_{3,3}(x) \\ &= (1 \cdot 2^0 \cdot 3^0 \cdot x_1^3 x_2^0 x_3^0 + 1 \cdot 2^0 \cdot 3^0 \cdot x_1^1 x_2^2 x_3^0 \\ &\quad -1 \cdot 2^2 \cdot 3^0 \cdot x_1^1 x_2^0 x_3^2)/\sqrt{5 \cdot 2^3 \cdot 3^0}, \end{aligned}$$

$$\begin{aligned} &\sqrt{(H^*_{3,4}, H^*_{3,4})^{-1}_{\mathrm{Hom}_3}}\, H^*_{3,4}(x) \\ &= (1 \cdot 2^0 \cdot 3^1 \cdot x_1^2 x_2^1 x_3^0 - 1 \cdot 2^0 \cdot 3^0 \cdot x_1^0 x_2^3 x_3^0)/\sqrt{1 \cdot 2^3 \cdot 3^1}, \end{aligned}$$

$$\begin{aligned} &\sqrt{(H^*_{3,5}, H^*_{3,5})^{-1}_{\mathrm{Hom}_3}}\, H^*_{3,5}(x) \\ &= (1 \cdot 2^0 \cdot 3^0 \cdot x_1^2 x_2^0 x_3^1 - 1 \cdot 2^0 \cdot 3^0 \cdot x_1^0 x_2^2 x_3^1)/\sqrt{1 \cdot 2^2 \cdot 3^0}, \end{aligned}$$

$$\sqrt{(H^*_{3,6}, H^*_{3,6})^{-1}_{\mathrm{Hom}_3}}\, H^*_{3,6}(x)$$

$$= (1 \cdot 2^0 \cdot 3^0 \cdot x_1^2 x_2^1 x_3^0 + 1 \cdot 2^0 \cdot 3^0 \cdot x_1^0 x_2^3 x_3^0 5 \cdot 2^3 \cdot 3^0 \cdot x_1^0 x_2^1 x_3^2)/\sqrt{1 \cdot 2^2 \cdot 3^0},$$

$$\sqrt{(H^*_{3,7}, H^*_{3,7})^{-1}_{\mathrm{Hom}_3}} H^*_{3,7}(x)$$
$$= (1 \cdot 2^0 \cdot 3^1 \cdot x_1^2 x_2^0 x_3^1 + 1 \cdot 2^0 \cdot 3^1 \cdot x_1^0 x_2^2 x_3^1$$
$$-1 \cdot 2^1 \cdot 3^0 \cdot x_1^0 x_2^0 x_3^3)/\sqrt{5 \cdot 2^2 \cdot 3^1}.$$

Therefore we obtain the system

$$H_{3,1}(x) = \frac{1}{\sqrt{24}}({x_1}^3 - 3x_1{x_2}^2), \tag{4.3.44}$$
$$H_{3,2}(x) = x_1 x_2 x_3, \tag{4.3.45}$$
$$H_{3,3}(x) = \frac{1}{\sqrt{40}}({x_1}^3 + x_1{x_2}^2 - 4x_1{x_3}^2), \tag{4.3.46}$$
$$H_{3,4}(x) = \frac{1}{\sqrt{24}}(3{x_1}^2 x_2 - {x_2}^3), \tag{4.3.47}$$
$$H_{3,5}(x) = \frac{1}{\sqrt{4}}({x_1}^2 x_3 - {x_2}^2 x_3), \tag{4.3.48}$$
$$H_{3,6}(x) = \frac{1}{\sqrt{40}}({x_1}^2 x_2 + {x_2}^3 - 4x_2 x_3^2), \tag{4.3.49}$$
$$H_{3,7}(x) = \frac{1}{\sqrt{60}}(3{x_1}^2 x_3 + 3{x_2}^2 x_3 - 2{x_3}^3) \tag{4.3.50}$$

of homogeneous harmonic polynomials of degree 3 (see Figure 4.3.1).
Finally, the orthonormal system of homogeneous harmonic polynomials of degree n (with respect to $(\cdot,\cdot)_{\mathcal{L}^2(\Omega)}$) is given as follows

$$\sqrt{\mu_3 (H^*_{3,m}, H^*_{3,m})^{-1}_{\mathrm{Hom}_3}}\; H^*_{3,m}, \quad m = 1 \;\ldots, 7 \tag{4.3.51}$$

with

$$\mu_3 = \frac{105}{4\pi} = \frac{1 \cdot 3 \cdot 5 \cdot 7}{4\pi}. \tag{4.3.52}$$

4.3.2 Generation of Linearly Independent Systems Via Recursion Relations

Our considerations have shown how a basis of $Harm_n$ can be computed entirely by integer operations from $2n+1$ systems of linear equations. The basis functions obtained can be orthonormalized exactly by means of the well-known Gram-Schmidt orthonormalization process. As result there are $2n+1$ homogeneous harmonic polynomials available (orthonormalized in the sense of $(\cdot,\cdot)_{Hom_n}$). But the disadvantage in that approach is that the linear systems of equations result in basis functions which are all involved in the computational work of the orthonormalization. This is the reason why [20] proposed an algorithm such that, for every degree $n \geq 2$, a basis of $Harm_n$ is computable exactly which divides itself in a natural way into four subsets such that the sets are mutually orthogonal and the intersection between any pair of subsets is empty.

The starting point for this approach is the fact that any homogeneous harmonic polynomial H_n in the three variables x_1, x_2, x_3 can be represented in the form (3.1.6), where $A_{n-j} : \mathbb{R}^3 \to \mathbb{R}$, $j = 0, \ldots, n$, are homogeneous polynomials of degree $n-j$ in the variables x_1, x_2, where, for $j = 0, ..., n-2$, A_{n-j-2} is determined recursively from A_{n-j} according to (3.1.7). In what follows we especially denote by $\mathcal{M}(n)$ the set of all multiindices $\alpha \in \mathbb{N}_0^2$ with $[\alpha] = \alpha_1 + \alpha_2 = n$:

$$\mathcal{M}(n) = \{(n-j, j), \quad j = 0, \ldots, n\}. \tag{4.3.53}$$

Obviously, $\mathcal{M}(n)$ consists of $n+1$ elements. Hence, corresponding to the set $\mathcal{M}(n)$, there exists an ordered system $A_{n,1}, \ldots, A_{n,n+1}$ of $n+1$ monomials $(x_1, x_2) \mapsto x_1^{\alpha_1} x_2^{\alpha_2}, \alpha_1 + \alpha_2 = n, (x_1, x_2)^T \in \mathbb{R}^2$. It is clear that the set of all linear combinations of $A_{n,j}, j = 1, ..., n+1$, spans the space of all homogeneous polynomials of degree n in $\mathbb{R}^2$. For brevity, we use $\mathcal{A}_n$ to denote the system of the $n+1$ monomials $\{A_{n,j}\}_{j=1,...,n+1}$.

Lemma 4.3.3. *The union $\mathcal{B}_n$ of the sets $\mathcal{B}_n^{(n-1)}$ and $\mathcal{B}_n^{(n)}$ given by*

$$\mathcal{B}_n^{(n-1)} = \left\{ \sum_{j=1}^{n}{}^{2}\, x_3^j A_{n-j,l}(x_1, x_2) \mid l = 1, ..., n \right\}, \tag{4.3.54}$$

$$\mathcal{B}_n^{(n)} = \left\{ \sum_{j=0}^{n}{}^{2}\, x_3^j A_{n-j,l}(x_1, x_2) \mid l = 1, ..., n+1 \right\} \tag{4.3.55}$$

with

$$A_{n-j-2,l}(x_1, x_2) = -\frac{1}{(j+1)(j+2)}\left(\left(\frac{\partial}{\partial x_1}\right)^2 + \left(\frac{\partial}{\partial x_2}\right)^2\right) A_{n-j,l}(x_1, x_2)$$

forms a basis of the linear space $Harm_n$, where $\sum^2$ means that every second summand will be omitted in the sum $\sum$.

Proof. All members of the subsystems $\mathcal{B}_n^{(n-1)}, \mathcal{B}_n^{(n)}$ are linearly independent, since the function $A_{n,j}$ are linearly independent. Both systems together possess $n + n + 1 = 2n + 1$ elements, i.e., they form a basis of $Harm_n$. □

Obviously,

$$H_{n,l}^{(n-1)}(x) = \sum_{j=1}^{n} 2\, x_3^j A_{n-j,l}(x_1, x_2) = \sum_{j=1}^{\lceil \frac{n}{2} \rceil} x_3^{2j-1} A_{n-2j+1,l}(x_1, x_2) \tag{4.3.56}$$

and

$$H_{n,l}^{(n)}(x) = \sum_{j=0}^{n} 2\, x_3^j A_{n-j,l}(x_1, x_2) = \sum_{j=0}^{\lfloor \frac{n}{2} \rfloor} x_3^{2j} A_{n-2j,l}(x_1, x_2), \tag{4.3.57}$$

where we have used the notation $\lceil \cdot \rceil$ and $\lfloor \cdot \rfloor$ for rounding real numbers $\lfloor t \rfloor = \max\{n \in \mathbb{Z} | n \leq t\}$, $\lceil t \rceil = \min\{n \in \mathbb{Z} | n \geq t\}, t \in \mathbb{R}$.

Next we verify the following theorem.

Theorem 4.3.4. *Let $\mathcal{B}_n^{(k)}$, $k = n-1$ and n, respectively, be defined as in Lemma 4.3.3 Then $\mathcal{B}_n^{(n-1)} \perp \mathcal{B}_n^{(n)}$ (in the sense of $(\cdot,\cdot)_{Hom_n}$).*

Proof. Since each $H_{n,l}^{(n-1)} \in \mathcal{B}_n^{(n-1)}$ is homogeneous, it can be represented in the form

$$\begin{aligned} H_{n,l}^{(n-1)}(x_1, x_2, x_3) &= \sum_{j=1}^{\lceil \frac{n}{2} \rceil} x_3^{2j-1} A_{n-2j+1}(x_1, x_2) \\ &= \sum_{j=1}^{\lceil \frac{n}{2} \rceil} x_3^{2j-1} \sum_{\substack{(\beta_1,\beta_2)^T \in \mathbb{N}_0^2, \\ \beta_1+\beta_2 = n-2j+1}} C_{\beta_1,\beta_2} x_1^{\beta_1} x_2^{\beta_2} \\ &= \sum_{j=1}^{\lceil \frac{n}{2} \rceil} \sum_{\substack{(\alpha_1,\alpha_2\alpha_3)^T \in \mathbb{N}_0^3 \\ \alpha_1+\alpha_2 = n-2j+1 \\ \alpha_3 = 2j-1}} C_{\alpha_1,\alpha_2,\alpha_3} x_1^{\alpha_1} x_2^{\alpha_2} x_3^{\alpha_3}. \end{aligned} \tag{4.3.58}$$

Analogously we obtain

$$H_{n,k}^{(n)}(x_1, x_2, x_3) = \sum_{i=0}^{\lfloor \frac{n}{2} \rfloor} \sum_{\substack{(\gamma_1,\gamma_2,\gamma_3)^T \in \mathbb{N}_0^3, \\ \gamma_1+\gamma_2 = n-2i, \\ \gamma_3 = 2i}} C_{\gamma_1,\gamma_2,\gamma_3} x_1^{\gamma_1} x_2^{\gamma_1} x_3^{\gamma_3}. \tag{4.3.59}$$

Thus it follows that

$$\left(H_{n,l}^{(n-1)}, H_{n,k}^{(n)}\right)_{\mathrm{Hom}_n} = \sum_{j=1}^{\lceil \frac{n}{2} \rceil} \sum_{i=0}^{\lfloor \frac{n}{2} \rfloor} \sum_{\substack{\alpha \in \mathbb{N}_0^3 \\ \alpha_1+\alpha_2 = n-2j+1 \\ \alpha_3 = 2j-1}} \sum_{\substack{\gamma \in \mathbb{N}_0^3 \\ \gamma_1+\gamma_2 = n-2i \\ \gamma_3 = 2i}} C_\alpha C_\gamma (\nabla_x)^\alpha x^\gamma \tag{4.3.60}$$

This completes the proof of Theorem 4.3.4. □

Theorem 4.3.4 shows that we already have obtained a basis of $Harm_n$ which is partially orthogonal. Thus we know that we have two subsystems which are orthogonal to each other. This can be improved by the following: The reason for the orthogonality of the subsystems is that one subsystem contains only polynomials with multiindices $\alpha = (\alpha_1, \alpha_2, \alpha_3)^T \in \mathbb{N}_0^3$, where α_3 is odd, whereas in the other subsystems are only polynomials with multiinidices $\alpha = (\alpha_1, \alpha_2, \alpha_3)^T \in \mathbb{N}_0^3$, where α_3 is even. This idea can be applied in the same way to the second component of the multiindices.

In more detail we let

$$\mathcal{B}_n = \bigcup_{k=n-1,n} \mathcal{B}_n^{(k)} \tag{4.3.61}$$

$$= \bigcup_{\substack{k=n-1,n \\ i=0,1}} \mathcal{B}_{n,i}^{(k)}, \tag{4.3.62}$$

where

$$B_{n,0}^{(k)} = \{H_k \in \mathcal{B}_n^{(k)} | \alpha_2 \text{ even}\}, \tag{4.3.63}$$

$$B_{n,1}^{(k)} = \{H_k \in \mathcal{B}_n^{(k)} | \alpha_2 \text{ odd}\}. \tag{4.3.64}$$

Let us denote by $\mathcal{M}_0(n)$, $\mathcal{M}_1(n)$ the sets of all multiindices given by

$$\begin{aligned} \mathcal{M}_0(n) &= \{(n-j, j) \mid j \text{ even},\ 0 \le j \le n\}, \\ \mathcal{M}_1(n) &= \{(n-j, j) \mid j \text{ odd},\ 0 \le j \le n\}. \end{aligned} \tag{4.3.65}$$

By virtue of this splitting of $\mathcal{M}(n)$, the set $\mathcal{A}_n$ defined above is separated into two subsystems $\mathcal{A}_{n,0}$, $\mathcal{A}_{n,1}$, namely

$$\mathcal{A}_{n,k} = \{(x_1, x_2) \to x_1^{\alpha_1} x_2^{\alpha_2},\ \alpha_1 + \alpha_2 = n,\ (\alpha_1, \alpha_2) \in \mathcal{M}_k(n)\} \tag{4.3.66}$$

for $k = 0, 1$.

Remark 4.3.2. *As subscripts for the sets $\mathcal{M}$ we use binary numbers. The binary digits reflect the odd/even pattern of the multiindices if we read 'even' for 0 and 'odd' for 1. We see that $\mathcal{M}_k(n)$ and $\mathcal{M}_k(n-1)$ have the same patterns in the second component, but the patterns differ in the first component.*

As required by Theorem 3.1.5, the Laplace operator $(\partial/\partial x_1)^2 + (\partial/\partial x_2)^2$ in $\mathbb{R}^2$ has to be applied repeatedly to a basis function $A_{n-1,l}$ or $A_{n,l}$ in order to generate the functions

$A_{n-j-2,l}$. Applying the Laplacian in $\mathbb{R}^2$ changes the multiindices, but does not change the odd/even pattern. Thus, a homogeneous harmonic polynomial generated by the elements of $\mathcal{A}_{n-1,k}$, $k = 0, 1$, can be represented as a sum of the kind

$$\sum_{\substack{[\alpha]=n,\ \alpha_3 odd \\ (\alpha_1,\alpha_2)\in\mathcal{M}_{(k)}(n-1)}} C_\alpha^{(n-1)} x_1^{\alpha_1} x_2^{\alpha_2} x_3^{\alpha_3},\ x = (x_1, x_2, x_3)^T, \tag{4.3.67}$$

while a homogeneous harmonic polynomial generated by the elements of $\mathcal{A}_{n,k}$, $k = 0, 1$, can be represented by a sum of the kind

$$\sum_{\substack{[\alpha]=n,\alpha_3 even \\ (\alpha_1,\alpha_2)\in\mathcal{M}_k(n)}} C_\alpha^{(n)} x_1^{\alpha_1} x_2^{\alpha_2} x_3^{\alpha_3},\ x = (x_1, x_2, x_3)^T. \tag{4.3.68}$$

Therefore we obtain

Theorem 4.3.5. *Let $\mathcal{B}_{n,k}^{(n-1)}$, $k = 0, 1$, be the set of homogeneous harmonic polynomials generated from the elements of $\mathcal{A}_{n-1,k}$. Moreover, let $\mathcal{B}_{n,k}^{(n)}$, $k = 0, 1$, be the sets of homogeneous harmonic polynomials generated from $\mathcal{A}_{n,k}$. Then*

$$\mathcal{B}_n = \mathcal{B}_{n,0}^{(n-1)} \cup \mathcal{B}_{n,1}^{(n-1)} \cup \mathcal{B}_{n,0}^{(n)} \cup \mathcal{B}_{n,1}^{(n)}$$

is a basis of $Harm_n$, where

$$\mathcal{B}_{n,0}^{(n-1)} \cup \mathcal{B}_{n,1}^{(n-1)} = \mathcal{B}_n^{(n-1)},\ \mathcal{B}_{n,0}^{(n-1)} \cap \mathcal{B}_{n,1}^{(n-1)} = \emptyset \tag{4.3.69}$$

$$\mathcal{B}_{n,0}^{(n)} \cup \mathcal{B}_{n,1}^{(n)} = \mathcal{B}_n^{(n)},\ \mathcal{B}_{n,0}^{(n)} \cap \mathcal{B}_{n,1}^{(n)} = \emptyset \tag{4.3.70}$$

and

$$\mathcal{B}_{n,k}^{(i)} \perp \mathcal{B}_{n,l}^{(j)}$$

(in the sense of $(\cdot,\cdot)_{Hom_n}$) for $(i,k) \neq (j,l)$; $n-1 \leq i,j \leq n$; $0 \leq k,l \leq 1$.

Generation of Orthonormal Systems

Applied to the functions of each subsystem $\mathcal{B}_{n,k}^{(i)}$; $i = n-1, n$; $k = 0, 1$; the Gram-Schmidt orthonormalization process can be performed exactly (over a finite subset of $\mathbb{Q}$) except for the final (square root) division of each polynomial by its norm. This yields an orthonormal basis of homogeneous harmonic polynomials in $(Harm_n, (\cdot,\cdot)_{Hom_n})$. To this orthonormal basis there corresponds an orthonormal basis in $(Harm_n, (\cdot,\cdot)_{\mathcal{L}^2(\Omega)})$ by multiplying each basis element with the factor $\sqrt{\mu_n}$.

In order to get an estimate of the amount of work saved if we orthonormalize the subsets rather than the whole set of basis functions, we have to know the number of functions in each subset $\mathcal{B}^{(i)}_{n,k}$; $i = n-1, n$; $k = 0, 1$, i.e., the number of elements in $\mathcal{M}_{(k)}(n-1)$ and $\mathcal{M}_{(k)}(n)$ for $k = 0, 1$. Let us define

$$\nu_k(n) = \#\mathcal{M}_{(k)}(n).$$

Then we easily see that for even n

$$\nu_k(n) = \begin{cases} \frac{n+2}{2}, & \text{if } k = 0, \\ \frac{n}{2}, & \text{if } k = 1, \end{cases}$$

while for odd n

$$\nu_k(n) = \begin{cases} \frac{n+1}{2}, & \text{if } k = 0, \\ \frac{n+1}{2}, & \text{if } k = 1. \end{cases}$$

Let us denote by $W_N(n)$ the amount of computational work to perform the Gram-Schmidt orthonormalization process for a subset of N homogeneous harmonic polynomials. In the i-th step are computed: i - scalar products (S), (i-1)- divisions (D); (rational/rational), (i-1)- multiplications (M); (rational x polynomial) and (i-1)- additions (A); (polynomial $\pm$ polynomial). The square root is not really performed. Thus $W_N(n)$ is equal to

$$\begin{aligned} W_N(n) &= \sum_{i=1}^{N} i \cdot S + (i-1)(D + M + A) \\ &= \frac{1}{2}(N(N+1)S + (N-1)N(D + M + A)) \\ &\approx \frac{1}{2}N^2(S + D + M + A) = \frac{1}{2}N^2 \cdot U, \end{aligned}$$

where

$$U = S + D + M + A.$$

For the orthonormalization of a set of $2n+1$ basis functions of $Harm_n$, i.e. for all functions of a basis of $Harm_n$,we get

$$W_{all}(n) = W_{2n+1}(n) \approx \tilde{W}_{2n+1}(n) = \frac{1}{2}(2n+1)^2 \cdot U.$$

The computational work for the individual orthonormalization of the four subsets is

$$\begin{aligned} W_{subsets}(n) &= \sum_{k=0}^{1} \sum_{j=n-1}^{n} W_{\nu_k(n)}(j) \\ \approx \tilde{W}_{subsets}(n) &= \frac{1}{2}U \sum_{k=0}^{1} \sum_{j=n-1}^{n} \nu_k^2(j). \end{aligned}$$

The amount of computational work which we save if we orthonormalize individually the basis subsets can be expressed by the ratio

$$\tilde{R}(n) = \frac{\tilde{W}_{all}(n)}{\tilde{W}_{subsets}(n)}.$$

We see that

$$\tilde{R}(n) < 4$$

and

$$\lim_{n\to\infty} \tilde{R}(n) = 4.$$

Even for very small n the limit 4 is nearly reached.
This means that the amount of computational work is reduced drastically when the elements of the basis have to be orthonormalized, because the orthonormalization process can be performed separately for the four individual subsets (whose numbers of elements on average is $(2n+1)/4$).
We start with the sets of multiindices

$$\begin{aligned}\mathcal{M}(4) &= \{(4,0),(3,1),(2,2),(1,3),(0,4)\},\\ \mathcal{M}(5) &= \{(5,0),(4,1),(3,2),(2,3),(1,4),(0,5)\}.\end{aligned}$$

They are split according to procedure described above into

$$\begin{aligned}\mathcal{M}_0(4) &= \{(4,0),(2,2),(0,4)\},\\ \mathcal{M}_1(4) &= \{(3,1),(1,3)\},\\ \mathcal{M}_0(5) &= \{(5,0),(3,2),(1,4)\},\\ \mathcal{M}_1(5) &= \{(4,1),(2,3),(0,5)\}.\end{aligned}$$

From each of these sets the corresponding homogeneous harmonic polynomials are derived. The polynomials read as follows:

$$\begin{aligned}\mathcal{M}_0(4):\ & 1x_1^4x_2^0x_3^1 - 2x_1^2x_2^0x_3^3 + \tfrac{1}{5}x_1^0x_2^0x_3^5,\\ & 1x_1^2x_2^2x_3^1 - \tfrac{1}{3}x_1^0x_2^2x_3^3 - \tfrac{1}{3}x_1^2x_2^0x_3^3 + \tfrac{1}{15}x_1^0x_2^0x_3^5,\\ & 1x_1^0x_2^4x_3^1 - 2x_1^0x_2^2x_3^3 + \tfrac{1}{5}x_1^0x_2^0x_3^5,\\ \mathcal{M}_1(4):\ & 1x_1^3x_2^1x_3^1 - 1x_1^1x_2^1x_3^3,\\ & 1x_1^1x_2^3x_3^1 - 1x_1^1x_2^1x_3^3,\end{aligned}$$

$$
\begin{aligned}
\mathcal{M}_0(5): \quad & 1x_1^5x_2^0x_3^0 - 10x_1^3x_2^0x_3^2 + 5x_1^1x_2^0x_3^4, \\
& 1x_1^3x_2^2x_3^0 - 3x_1^1x_2^2x_3^2 - 1x_1^3x_2^0x_3^2 + 1x_1^1x_2^0x_3^4, \\
& 1x_1^1x_2^4x_3^0 - 6x_1^1x_2^2x_3^2 + 1x_1^1x_2^0x_3^4, \\
\mathcal{M}_1(5): \quad & 1x_1^4x_2^1x_3^0 - 6x_1^2x_2^1x_3^2 + 1x_1^0x_2^1x_3^4, \\
& 1x_1^2x_2^3x_3^0 - 1x_1^0x_2^3x_3^2 - 3x_1^2x_2^1x_3^2 + 1x_1^0x_2^1x_3^4.
\end{aligned}
$$

An illustration of members of the system of homogeneous harmonic polynomials for degree 5 is given in Figure 4.3.2.

For the $(\cdot,\cdot)_{\mathrm{H}om_5}$-orthonormalized set of polynomials we find the following 11 monomials:

$$
\begin{aligned}
\mathcal{M}_0(4): \quad & \sqrt{\tfrac{5}{384}}(1x_1^4x_2^0x_3^1 - 2x_1^2x_2^0x_3^3 + \tfrac{1}{5}x_1^0x_2^0x_3^5), \\
& \tfrac{1}{\sqrt{6}}(1x_1^2x_2^2x_3^1 - \tfrac{1}{3}x_1^0x_2^0x_3^3 - \tfrac{1}{12}x_1^2x_2^0x_3^3 + \tfrac{1}{24}x_1^0x_2^0x_3^5 \\
& -\tfrac{1}{8}x_1^4x_2^0x_3^1), \\
& \tfrac{1}{\sqrt{63}}(1x_1^0x_2^4x_3^1 - \tfrac{3}{2}x_1^0x_2^2x_3^3 + \tfrac{1}{8}x_1^0x_2^0x_3^5 + \tfrac{1}{8}x_1^4x_2^0x_3^1 \\
& +\tfrac{1}{4}x_1^2x_2^0x_3^3 - \tfrac{3}{2}x_1^1x_2^2x_3^1), \\
\mathcal{M}_1(4): \quad & \tfrac{1}{\sqrt{12}}(1\cdot x_1^3x_2^1x_3^1 - 1x_1^1x_2^1x_3^3), \\
& \tfrac{1}{\sqrt{9}}(1x_1^1x_2^3x_3^1 - \tfrac{1}{2}x_1^1x_2^1x_3^3 - \tfrac{1}{2}x_1^3x_2^1x_3^1), \\
\mathcal{M}_0(5): \quad & \tfrac{1}{\sqrt{1920}}(1x_1^5x_2^0x_3^0 - 10x_1^3x_2^0x_3^2 + 5x_1x_2^0x_3^4), \\
& \tfrac{1}{\sqrt{54}}(1x_1^3x_2^2x_3^0 - 3x_1^1x_2^2x_3^2 + \tfrac{1}{4}x_1^3x_2^0x_3^2 \\
& \qquad +\tfrac{1}{4}x_1^3x_2^0x_3^4 + \tfrac{3}{8}x_1^1x_2^0x_3^4 - \tfrac{1}{8}x_1^5x_2^0x_3^0), \\
& \tfrac{1}{\sqrt{63}}(1x_1^1x_2^4x_3^0 - \tfrac{3}{2}x_1^1x_2^2x_3^2 + \tfrac{1}{8}x_1^1x_2^0x_3^4 \\
& \qquad +\tfrac{1}{8}x_1^5x_2^0x_3^0 + \tfrac{1}{4}x_1^3x_2^0x_3^2 - \tfrac{3}{2}x_1^3x_1^2x_3^0), \\
\mathcal{M}_1(5): \quad & \tfrac{1}{\sqrt{192}}(1x_1^4x_2^1x_3^0 - 6x_1^2x_2^1x_3^2 + 1x_1^0x_2^1x_3^4), \\
& \tfrac{1}{\sqrt{36}}(1x_1^2x_3^3x_3^0 - 1x_1^0x_2^3x_3^2 + \tfrac{1}{2}x_1^0x_2^1x_3^4), \\
& \tfrac{1}{\sqrt{945}}\left(1x_1^0x_2^5x_3^0 - 5x_1^0x_2^3x_3^2 + \tfrac{15}{8}x_2^0x_2^1x_3^4\right. \\
& \qquad \left.+\tfrac{15}{8}x_1^4x_2^1x_3^0 + \tfrac{15}{4}x_1^2x_2^1x_3^2 - 5x_1^2x_2^3x_3^0\right).
\end{aligned}
$$

An illustration of members of the $(\cdot,\cdot)_{\mathrm{H}om_n}$-orthonormalized system of homogeneous harmonic polynomials for degree 5 is given in Figure 4.3.3.

4.3.3 Generation of Scalar Spherical Harmonics and Scalar Outer Harmonics

One more step is needed to produce a $(\cdot,\cdot)_{\mathcal{L}^2(\Omega)}$-orthonormalized set of polynomials from the above given polynomials. By multiplying the $(\cdot,\cdot)_{\mathrm{Hom}_n}$-orthonormalized system with the normalizing factor $\sqrt{\mu_n}$, where μ_n is defined by (3.1.13) we obtain a $(\cdot,\cdot)_{\mathcal{L}^2(\Omega)}$-orthonormalized set of polynomials. Therefore, let $\{H_{n,m}\}_{m=1,\ldots,2n+1}$ be a set of $(\cdot,\cdot)_{\mathrm{Hom}_n}$-orthonormalized polynomials in $Harm_n$. Then, with $x = |x|\xi$, $\xi \in \Omega$ and μ_n defined by (3.1.13) the system $\{Y_{n,m}\}_{m=1,\ldots,2n+1}$ with

$$Y_{n,m}(\xi) = \sqrt{\mu_n} H_{n,m}\left(\frac{x}{|x|}\right), \quad x \in \Omega, \tag{4.3.71}$$

is a set of $(\cdot,\cdot)_{\mathcal{L}^2(\Omega)}$-orthonormalized polynomials in $Harm_n$.
In our above example, for the case of $n = 5$, we obtain $\mu_n = \frac{10395}{4\pi}$. An illustration of members of the $(\cdot,\cdot)_{\mathcal{L}^2(\Omega)}$-orthonormalized system of homogeneous harmonic polynomials for degree 5 is given in Figure 4.3.4.

Let us now provide further numerical examples for the generation of the extension of spherical harmonics to arbitrary spheres and finally to scalar outer harmonics.
The extension to an arbitrary sphere with radius R can be computed (due to (4.1.1)) by multiplying each member of the $(\cdot,\cdot)_{\mathcal{L}^2(\Omega)}$-orthonormalized system of homogeneous harmonic polynomials, given by (4.3.71), with $\frac{1}{R}$, i.e., for $R > 0$,

$$Y^R_{n,m}(x) = \frac{1}{R} Y_{n,m}\left(\frac{x}{|x|}\right), \quad x \in \Omega_R. \tag{4.3.72}$$

An example is given in Figure 4.3.5 for the extension from a sphere with $R_1 = 1$ to a sphere with $R_2 = 1.3$ (only a part of the spherical harmonics is extended for a better visualization).
We proceed in a similar way to obtain scalar outer harmonics (4.2.1). Therefore we use the result of the extended set of $(\cdot,\cdot)_{\mathcal{L}^2(\Omega)}$-orthonormalized polynomials, given by (4.3.72) and multiply with the factor $\left(\frac{R}{|x|}\right)^{n+1}$, $x \in \mathbb{R}^3 \setminus \{0\}$, to obtain the set of scalar outer harmonics $\{H^R_{n,m}\}_{\substack{n\in\mathbb{N}_0;\\ m=1,\ldots,2n+1}}$, i.e.,

$$H^R_{n,m}(x) = \left(\frac{R}{|x|}\right)^{n+1} Y^R_{n,m}(x), \quad x \in \mathbb{R}^3 \setminus \{0\}.$$

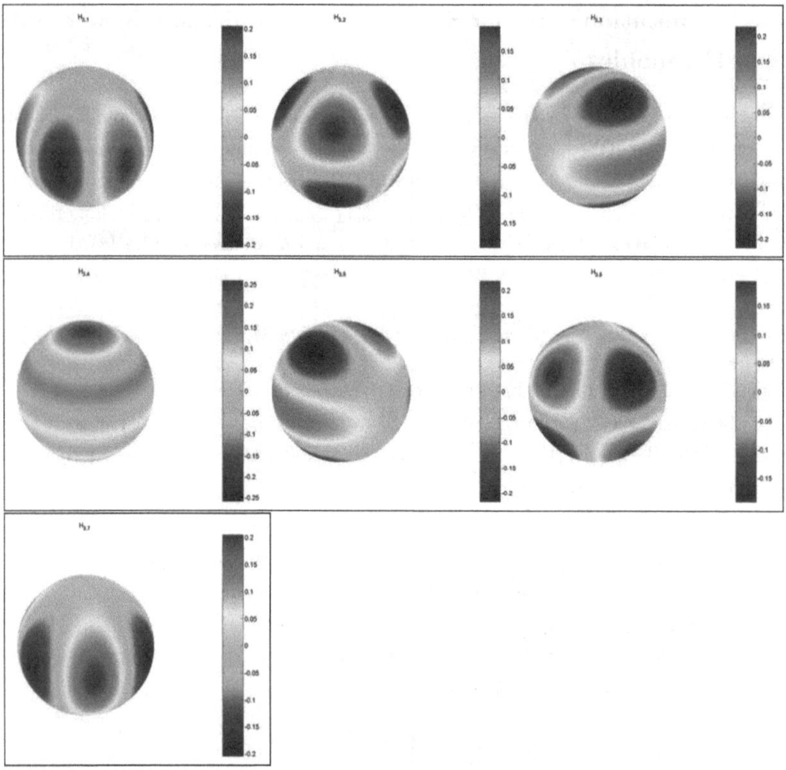

Figure 4.3.1: Illustration of the $(\cdot,\cdot)_{\mathrm{Hom}_3}$−orthonormalized system of homogeneous harmonic polynomials on the unit sphere.

Figure 4.3.6 provides some examples of scalar outer harmonics on a regular surface for degree 5 and order $m = 5, 6, 11$.

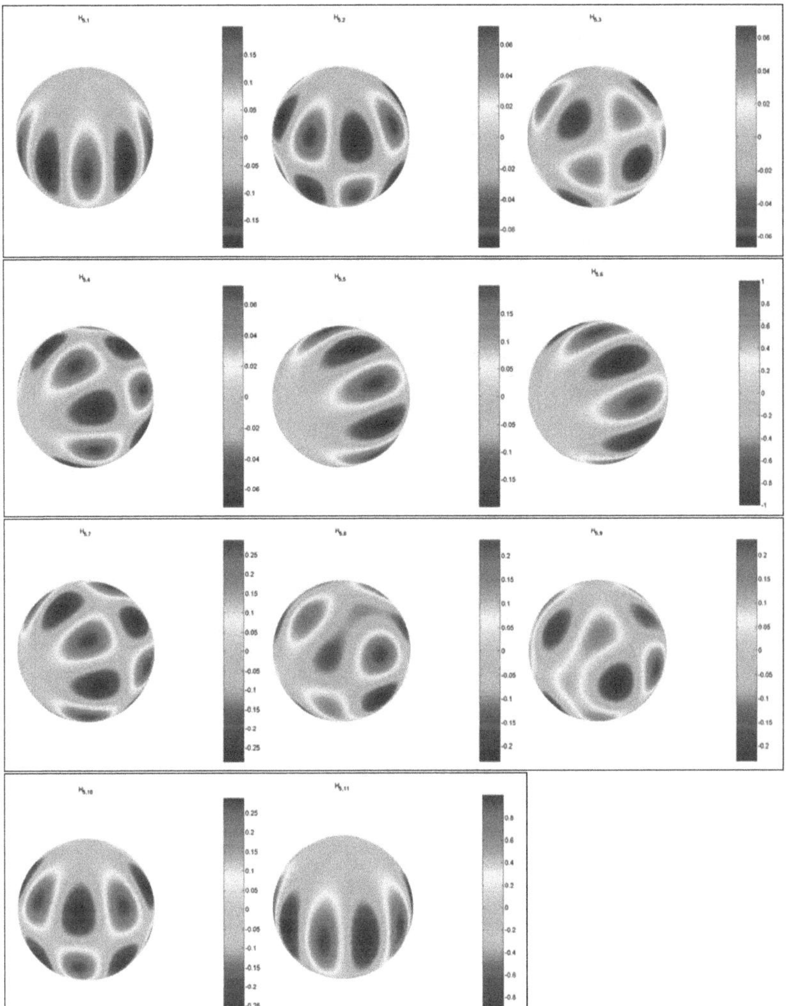

Figure 4.3.2: Members of the system of homogeneous harmonic polynomials for degree 5 and order $m = 1, \ldots, 11$.

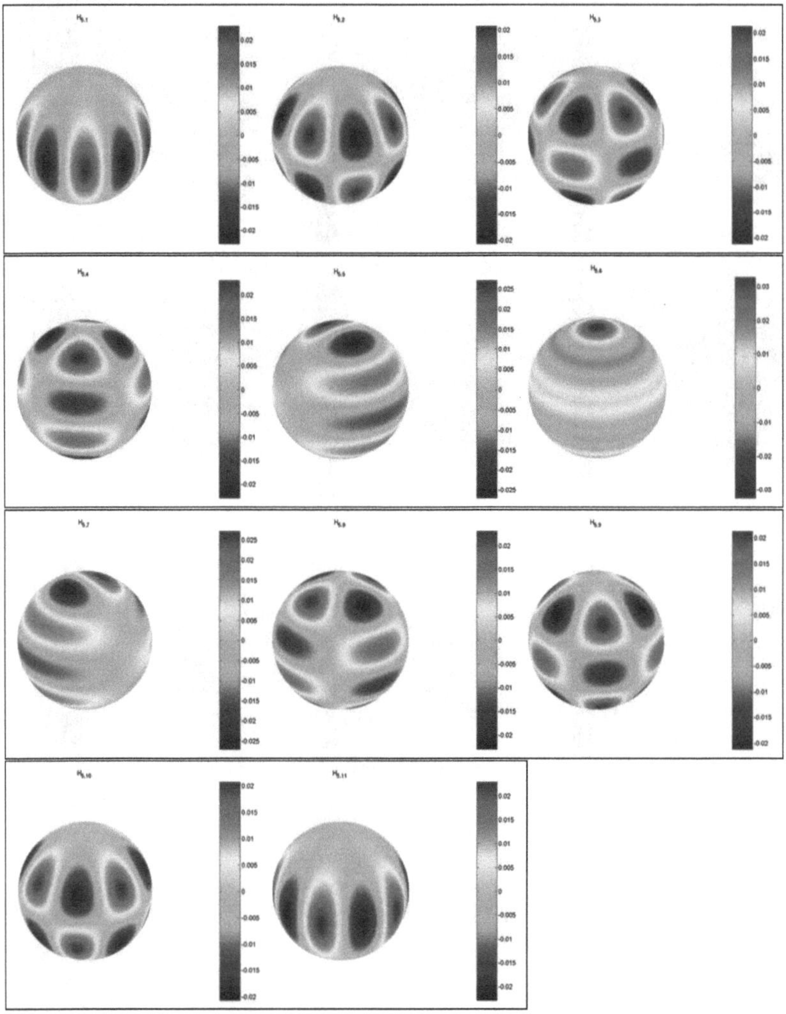

Figure 4.3.3: Members of the $(\cdot,\cdot)_{\mathrm{Hom}_n}$−orthonormalized system of homogeneous harmonic polynomials for degree 5 and order $m = 1, \ldots, 11$.

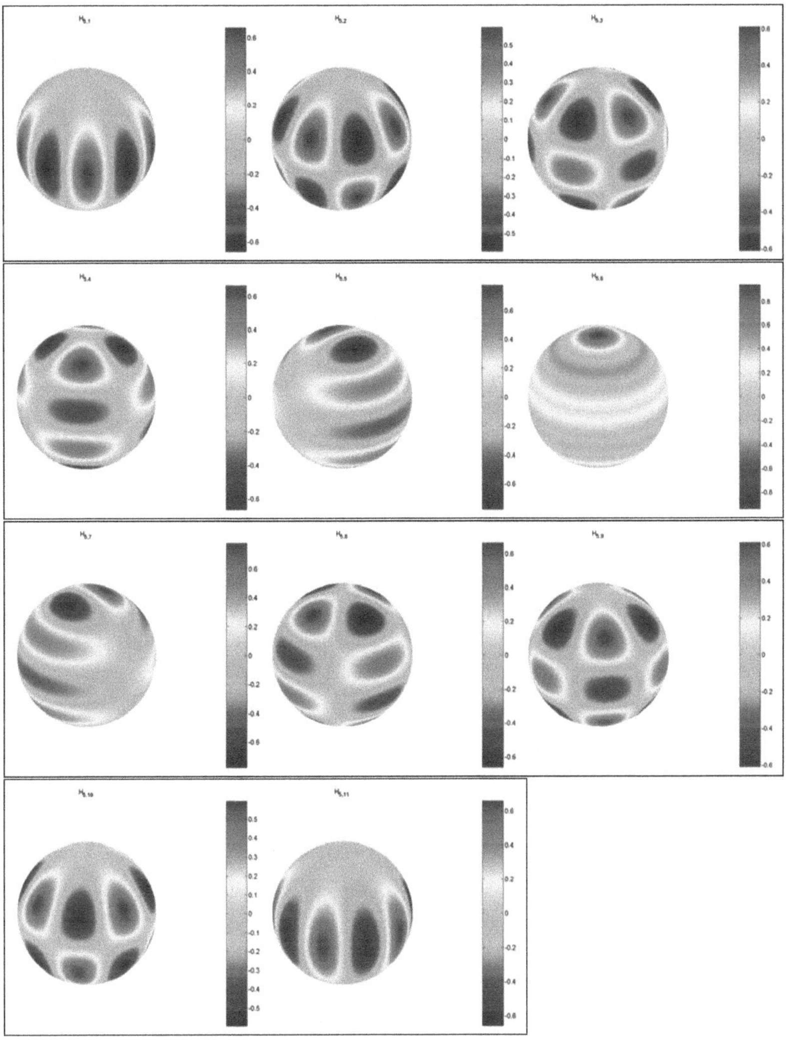

Figure 4.3.4: Members of the $(\cdot,\cdot)_{\mathcal{L}^2(\Omega)}$–orthonormalized system of homogeneous harmonic polynomials for degree 5 and order $m = 1, \ldots, 11$.

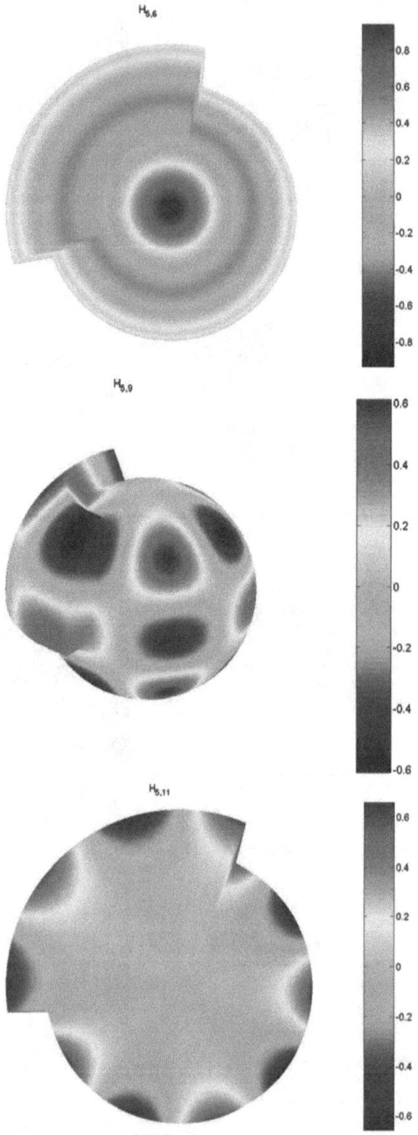

Figure 4.3.5: Illustration where scalar spherical harmonics have been extended for a part of the sphere (for better illustration) from $R_1 = 1$ to $R_2 = 1.3$ for degree 5 and order $m = 6, 9, 11$ (from top to bottom).

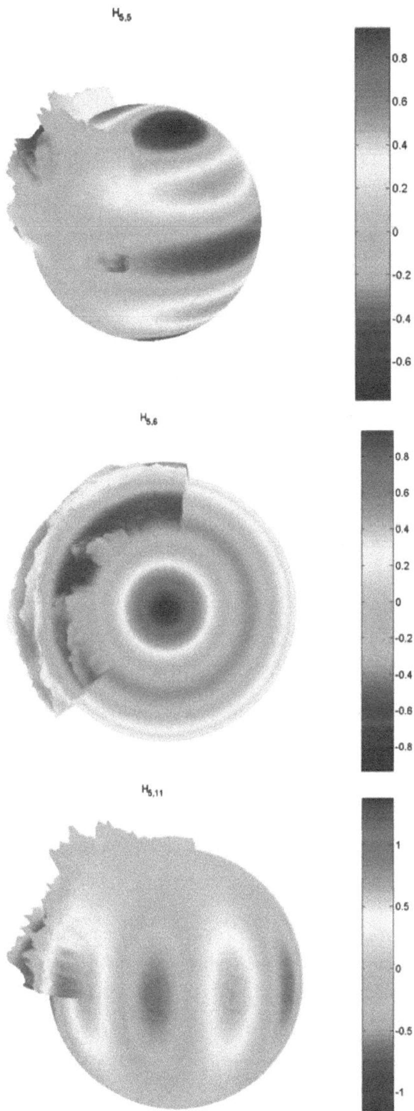

Figure 4.3.6: Illustration of scalar outer harmonics where a part of the sphere has been replaced by a regular surface (for better illustration) for degree 5 and order $m = 5, 6, 11$ (from top to bottom).

4.4 Exact Generation of Vector Spherical Harmonics and Vector Outer Harmonics

In what follows we present a way for determining $l^2(\Omega)-$orthonormal systems of vector spherical harmonics in terms of cartesian coordinates, avoiding problems arising from singularities of a spherical coordinate system.
Let $H_{n,m}$, $m = 1, \ldots, 2n+1$, be a Hom_n-orthonormal system of homogeneous harmonic polynomials of degree n, represented by

$$H_{n,m}(x) = \sum_{[\alpha]=n} C_\alpha^m x^\alpha \tag{4.4.1}$$

with known real members C_α^m (as defined in (3.1.1)).

Then we know that $Y_{n,m}(\xi) = \sqrt{\mu_n} H_{n,m}(\xi)$, $\xi \in \Omega$, μ_n given by (3.1.13) constitutes an $\mathcal{L}^2(\Omega)$-orthogonal system of spherical harmonics. Therefore, via the well-known procedure, by letting

$$y_{n,m}^{(i)} = (\mu_n^{(i)})^{1/2} o^{(i)} Y_{n,m};\ m = 1, \ldots, 2n+1,\ i \in 1, 2, 3,$$

an $l^2(\Omega)$-orthonormal system of vector spherical harmonics of kind i is found.

More explicitly,

$$\begin{aligned} y_{n,m}^{(1)}(\xi) &= \sqrt{\mu_n}\, \mathrm{h}_{n,m}^{(1)}(x)|_{|x|=1} = \sqrt{\mu_n}\, \mathrm{h}_{n,m}^{(1)}(\xi), \\ y_{n,m}^{(2)}(\xi) &= \sqrt{\mu_n}(n(n+1))^{-1/2}\, \mathrm{h}_{n,m}^{(2)}(x)|_{|x|=1} = \sqrt{\mu_n}(n(n+1))^{-1/2}\, \mathrm{h}_{n,m}^{(2)}(\xi), \\ y_{n,m}^{(3)}(\xi) &= \sqrt{\mu_n}(n(n+1))^{-1/2}\, \mathrm{h}_{n,m}^{(3)}(x)|_{|x|=1} = \sqrt{\mu_n}(n(n+1))^{-1/2}\, \mathrm{h}_{n,m}^{(3)}(\xi), \end{aligned}$$

where

$$\begin{aligned} \mathrm{h}_{n,m}^{(1)}(x) &= H_{n,m}(x)x, \\ \mathrm{h}_{n,m}^{(2)}(x) &= x^2 \nabla_x H_{n,m}(x) - nH_{n,m}(x)x, \\ \mathrm{h}_{n,m}^{(3)}(x) &= x \wedge \nabla_x H_{n,m}(x), \end{aligned}$$

with $x = r\xi$, $r = |x|$, $\xi \in \Omega$, and $\nabla^* = lim_{r\to 1}(r\nabla_x - x\frac{\partial}{\partial r})$. Our purpose is to determine the vector spherical harmonics with exact integer arithmetic. We base our considerations on the representation

$$\mathrm{h}_{n,m}^{(i)}(x) = \sum_{k=1}^{3} \varepsilon^k \left(\sum_{[\alpha]=l_i} D_{\alpha;m}^{i,k} x^\alpha \right), \tag{4.4.2}$$

where we have used the abbreviations $l_1 = l_2 = n+1, l_3 = n$. Using the already known identities (3.3.17)-(3.3.19),

$$y_{n,m}^{(1)}(\xi) = \sqrt{\frac{\mu_n}{\mu_n^{(1)}}}\frac{1}{2n+1}(k_n^{(1)} + k_n^{(2)})H_{n,m}(r\xi)|_{r=1}, \tag{4.4.3}$$

$$y_{n,m}^{(2)}(\xi) = \sqrt{\frac{\mu_n}{\mu_n^{(2)}}}\frac{1}{2n+1}(-nk_n^{(1)} + (n+1)k_n^{(2)})H_{n,m}(r\xi)|_{r=1}, \tag{4.4.4}$$

$$y_{n,m}^{(3)}(\xi) = \sqrt{\frac{\mu_n}{\mu_n^{(3)}}}k_n^{(3)}H_{n,m}(r\xi)|_{r=1} \tag{4.4.5}$$

we find by observing the definition of the $k_n^{(i)}$-operators (see Definition 3.3.3)

$$y_{n,m}^{(1)}(\xi) = \sqrt{\frac{\mu_n}{\mu_n^{(1)}}}\sum_{[\alpha]=n} C_\alpha^m \begin{pmatrix} \xi_1^{\alpha_1+1}\xi_2^{\alpha_2}\xi_3^{\alpha_3} \\ \xi_1^{\alpha_1}\xi_2^{\alpha_2+1}\xi_3^{\alpha_3} \\ \xi_1^{\alpha_1}\xi_2^{\alpha_2}\xi_3^{\alpha_3+1} \end{pmatrix},$$

$$y_{n,m}^{(2)}(\xi) = \sqrt{\frac{\mu_n}{\mu_n^{(2)}}}\sum_{[\alpha]=n} C_\alpha^m \begin{pmatrix} \alpha_1\xi_1^{\alpha_1-1}\xi_2^{\alpha_2}\xi_3^{\alpha_3}(\xi_1^2+\xi_2^2+\xi_3^2) - n\xi_1^{\alpha_1+1}\xi_2^{\alpha_2}\xi_3^{\alpha_3} \\ \alpha_2\xi_1^{\alpha_1}\xi_2^{\alpha_2-1}\xi_3^{\alpha_3}(\xi_1^2+\xi_2^2+\xi_3^2) - n\xi_1^{\alpha_1}\xi_2^{\alpha_2+1}\xi_3^{\alpha_3} \\ \alpha_3\xi_1^{\alpha_1}\xi_2^{\alpha_2}\xi_3^{\alpha_3-1}(\xi_1^2+\xi_2^2+\xi_3^2) - n\xi_1^{\alpha_1}\xi_2^{\alpha_2}\xi_3^{\alpha_3+1} \end{pmatrix},$$

$$y_{n,m}^{(3)}(\xi) = \sqrt{\frac{\mu_n}{\mu_n^{(3)}}}\sum_{[\alpha]=n} C_\alpha^m \begin{pmatrix} \alpha_3\xi_1^{\alpha_1}\xi_2^{\alpha_2+1}\xi_3^{\alpha_3-1} - \alpha_2\xi_1^{\alpha_1}\xi_2^{\alpha_2-1}\xi_3^{\alpha_3+1} \\ \alpha_1\xi_1^{\alpha_1-1}\xi_2^{\alpha_2}\xi_3^{\alpha_3+1} - \alpha_3\xi_1^{\alpha_1+1}\xi_2^{\alpha_2}\xi_3^{\alpha_3-1} \\ \alpha_2\xi_1^{\alpha_1+1}\xi_2^{\alpha_2-1}\xi_3^{\alpha_3} - \alpha_1\xi_1^{\alpha_1-1}\xi_2^{\alpha_2+1}\xi_3^{\alpha_3} \end{pmatrix}.$$

Hence, the coefficients $D_{\beta,m}^{i,k}$ occuring in Equation (4.4.2) are found. We organize their computation by a matrix-matrix multiplication:

$$D_{\beta;m}^{i,k} = \sqrt{\frac{\mu_n}{\mu_n^{(i)}}}\sum_{[\alpha]=n} C_\alpha^m M_{\beta\alpha;m}^{i,k}$$

for $i,k = 1,2,3$; $m = 1,\ldots,2n+1$; $[\beta] = n+1$ in the cases $i = 1,2$ respectively $[\beta] = n$ if $i = 3$. The matrices $\mathbf{m}^{i,k} = (M_{\beta,\alpha}^{i,k})$ have $\binom{n+2}{2}$ rows and $\binom{n}{2}$ columns in the cases $i = 1,2$ and they have $\binom{n}{2}$ rows and columns if $i = 3$. It is easy to develop that

$$M_{\beta\alpha;m}^{1,1} = \begin{cases} 1, & \text{if } \beta-\alpha = (1,0,0)^T, \\ 0, & \text{otherwise,} \end{cases}$$

$$M_{\beta\alpha;m}^{1,2} = \begin{cases} 1, & \text{if } \beta-\alpha = (0,1,0)^T, \\ 0, & \text{otherwise,} \end{cases}$$

$$M^{1,3}_{\beta\alpha;m} = \begin{cases} 1, & \text{if } \beta-\alpha=(0,0,1)^T, \\ 0, & \text{otherwise}, \end{cases}$$

$$M^{2,1}_{\beta\alpha;m} = \begin{cases} \alpha_1-n, & \text{if } \beta-\alpha=(1,0,0)^T, \\ \alpha_1, & \text{if } \beta-\alpha\in\{(-1,2,0)^T,(-1,0,2)^T\}, \\ 0, & \text{otherwise}, \end{cases}$$

$$M^{2,2}_{\beta\alpha;m} = \begin{cases} \alpha_2-n, & \text{if } \beta-\alpha=(0,1,0)^T, \\ \alpha_2, & \text{if } \beta-\alpha\in\{(2,-1,0)^T,(0,-1,2)^T\}, \\ 0, & \text{otherwise}, \end{cases}$$

$$M^{2,3}_{\beta\alpha;m} = \begin{cases} \alpha_3-n, & \text{if } \beta-\alpha=(0,0,1)^T, \\ \alpha_3, & \text{if } \beta-\alpha=\{(2,0,-1)^T,(0,2,-1)^T\}, \\ 0, & \text{otherwise}, \end{cases}$$

$$M^{3,1}_{\beta\alpha;m} = \begin{cases} \alpha_3, & \text{if } \beta-\alpha=(0,1,-1)^T, \\ -\alpha_2, & \text{if } \beta-\alpha=(0,-1,1)^T, \\ 0, & \text{otherwise}, \end{cases}$$

$$M^{3,2}_{\beta\alpha;m} = \begin{cases} \alpha_1, & \text{if } \beta-\alpha=(-1,0,1)^T, \\ -\alpha_3, & \text{if } \beta-\alpha=(1,0,-1)^T, \\ 0, & \text{otherwise}, \end{cases}$$

$$M^{3,3}_{\beta\alpha;m} = \begin{cases} \alpha_2, & \text{if } \beta-\alpha=(1,-1,0)^T, \\ -\alpha_1, & \text{if } \beta-\alpha=(-1,1,0)^T, \\ 0, & \text{otherwise}. \end{cases}$$

As a starting point for our example, we consider for $n = 3$ the $(\cdot,\cdot)_{Hom_3}$-orthonormal system given by (4.3.44)-(4.3.50) of homogeneous harmonic polynomials of degree 3 (see Figure 4.3.1). Then we obtain with $\sqrt{\mu_n} = \sqrt{\frac{105}{4\pi}}$

$$y^{(1)}_{3,1}(\xi) = \sqrt{\frac{105}{4\pi}}\frac{1}{\sqrt{24}}({\xi_1}^3 - 3\xi_1{\xi_2}^2)\xi,$$

$$y^{(1)}_{3,2}(\xi) = \sqrt{\frac{105}{4\pi}}\xi_1\xi_2\xi_3\xi,$$

$$y^{(1)}_{3,3}(\xi) = \sqrt{\frac{105}{4\pi}}\frac{1}{\sqrt{40}}({\xi_1}^3 + \xi_1{\xi_2}^2 - 4\xi_1{\xi_3}^2)\xi,$$

$$y^{(1)}_{3,4}(\xi) = \sqrt{\frac{105}{4\pi}}\frac{1}{\sqrt{24}}(3{\xi_1}^2\xi_2 - {\xi_2}^3)\xi,$$

$$
\begin{aligned}
y_{3,5}^{(1)}(\xi) &= \sqrt{\frac{105}{4\pi}}\frac{1}{\sqrt{4}}({\xi_1}^2\xi_3 - {\xi_2}^2\xi_3)\xi, \\
y_{3,6}^{(1)}(\xi) &= \sqrt{\frac{105}{4\pi}}\frac{1}{\sqrt{40}}({\xi_1}^2\xi_2 + {\xi_2}^3 - 4\xi_2\xi_3^2)\xi, \\
y_{3,7}^{(1)}(\xi) &= \sqrt{\frac{105}{4\pi}}\frac{1}{\sqrt{60}}(3{\xi_1}^2\xi_3 + 3{\xi_2}^2\xi_3 - 2{\xi_3}^3)\xi,
\end{aligned}
$$

and

$$
\begin{aligned}
y_{3,1}^{(2)}(\xi) &= \sqrt{\frac{105}{28\pi}}\frac{1}{\sqrt{24}}\begin{pmatrix} -3{\xi_1}^2({\xi_1}^2 - 3{\xi_2}^2) + 3({\xi_1}^2 - {\xi_2}^2) \\ -3\xi_1\xi_2({\xi_1}^2 - 3{\xi_2}^2) - 6\xi_1\xi_2 \\ -3\xi_1\xi_3({\xi_1}^2 - 3{\xi_2}^2) \end{pmatrix}, \\
y_{3,2}^{(2)}(\xi) &= \sqrt{\frac{105}{28\pi}}\begin{pmatrix} -3{\xi_1}^2\xi_2\xi_3 + \xi_2\xi_3 \\ -3\xi_1{\xi_2}^2\xi_3 + \xi_1\xi_3 \\ -3\xi_1\xi_2{\xi_3}^2 + \xi_1\xi_2 \end{pmatrix}, \\
y_{3,3}^{(2)}(\xi) &= \sqrt{\frac{105}{28\pi}}\frac{1}{\sqrt{40}}\begin{pmatrix} -3{\xi_1}^2(1 - 5{\xi_3}^2) + 3{\xi_1}^2 + {\xi_2}^2 - 4{\xi_3}^2 \\ -3\xi_1\xi_2(1 - 5{\xi_3}^2) + 2\xi_1\xi_2 \\ -3\xi_1\xi_3(1 - 5{\xi_3}^?) - 8\xi_1\xi_3 \end{pmatrix}, \\
y_{3,4}^{(2)}(\xi) &= \sqrt{\frac{105}{28\pi}}\frac{1}{\sqrt{24}}\begin{pmatrix} -3\xi_1\xi_2(3{\xi_1}^2 - {\xi_2}^2) + 6\xi_1\xi_2 \\ -3{\xi_2}^2(3{\xi_1}^2 - {\xi_2}^2) + 3{\xi_1}^2 - 3{\xi_2}^2 \\ -3\xi_2\xi_3(3{\xi_1}^2 - {\xi_2}^2) \end{pmatrix}, \\
y_{3,5}^{(2)}(\xi) &= \sqrt{\frac{105}{28\pi}}\frac{1}{\sqrt{4}}\begin{pmatrix} -3\xi_1\xi_3({\xi_1}^2 - {\xi_2}^2) + 2\xi_1\xi_3 \\ -3\xi_2\xi_3({\xi_1}^2 - {\xi_2}^2) - 2\xi_2\xi_3 \\ -3\xi_3({\xi_1}^2 - {\xi_2}^2) + {\xi_1}^2 - {\xi_2}^2 \end{pmatrix}, \\
y_{3,6}^{(2)}(\xi) &= \sqrt{\frac{105}{28\pi}}\frac{1}{\sqrt{40}}\begin{pmatrix} -3\xi_1\xi_2(1 - 5\xi_3^2) + 2\xi_1\xi_2 \\ -3{\xi_2}^2(1 - 5\xi_3^2) + {\xi_1}^2 + 3{\xi_2}^2 - 4{\xi_3}^2 \\ -3\xi_2\xi_3(1 - 5\xi_3^2) - 8\xi_2\xi_3 \end{pmatrix}, \\
y_{3,7}^{(2)}(\xi) &= \sqrt{\frac{105}{28\pi}}\frac{1}{\sqrt{60}}\begin{pmatrix} -3\xi_1\xi_3(3 - 5{\xi_3}^2) + 6\xi_1\xi_3 \\ -3\xi_2\xi_3(3 - 5{\xi_3}^2) + 6\xi_2\xi_3 \\ -3{\xi_3}^2(3 - 5{\xi_3}^2) + 3(1 - 3{\xi_3}^2) \end{pmatrix},
\end{aligned}
$$

and

$$
y_{3,1}^{(3)}(\xi) = \sqrt{\frac{105}{28\pi}}\frac{1}{\sqrt{24}}\xi \wedge \begin{pmatrix} 3({\xi_1}^2 - {\xi_2}^2) \\ -6\xi_1\xi_2 \\ 0 \end{pmatrix},
$$

$$y_{3,2}^{(3)}(\xi) = \sqrt{\frac{105}{28\pi}}\xi \wedge \begin{pmatrix} \xi_2\xi_3 \\ \xi_1\xi_3 \\ \xi_1\xi_2 \end{pmatrix},$$

$$y_{3,3}^{(3)}(\xi) = \sqrt{\frac{105}{28\pi}}\frac{1}{\sqrt{40}}\xi \wedge \begin{pmatrix} 3{\xi_1}^2 + {\xi_2}^2 - 4{\xi_3}^2 \\ 2\xi_1\xi_2 \\ -8\xi_1\xi_3 \end{pmatrix},$$

$$y_{3,4}^{(3)}(\xi) = \sqrt{\frac{105}{28\pi}}\frac{1}{\sqrt{24}}\xi \wedge \begin{pmatrix} 6\xi_1\xi_2 \\ 3({\xi_1}^2 - {\xi_2}^2) \\ 0 \end{pmatrix},$$

$$y_{3,5}^{(3)}(\xi) = \sqrt{\frac{105}{28\pi}}\frac{1}{\sqrt{4}}\xi \wedge \begin{pmatrix} 2\xi_1\xi_3 \\ -2\xi_2\xi_3 \\ {\xi_1}^2 - {\xi_2}^2 \end{pmatrix},$$

$$y_{3,6}^{(3)}(\xi) = \sqrt{\frac{105}{28\pi}}\frac{1}{\sqrt{40}}\xi \wedge \begin{pmatrix} 2\xi_1\xi_2 \\ {\xi_1}^2 + 3{\xi_2}^2 - 4{\xi_3}^2 \\ -8\xi_2\xi_3 \end{pmatrix},$$

$$y_{3,7}^{(3)}(\xi) = \sqrt{\frac{105}{28\pi}}\frac{1}{\sqrt{60}}\xi \wedge \begin{pmatrix} 6\xi_1\xi_3 \\ 6\xi_2\xi_3 \\ 3(1 - 3{\xi_3}^2) \end{pmatrix}.$$

The set of vector spherical harmonics $y_{3,m}^{(i)}$, $i = 1, 2, 3$; $m = 1, \ldots, 2n + 1$ is illustrated in Figures 4.4.1- 4.4.2. Further, some examples of the vector spherical harmonics $y_{10,15}^{(i)}$, $i = 1, 2, 3$ are shown in Figures 4.4.3- 4.4.5. The vector spherical harmonics of type 1 represent a radial field while the vector spherical harmonics of type 2 and 3 are tangential fields. These properties can also be seen in Figures 4.4.3- 4.4.5.

To derive the system of vector spherical harmonics $\{u_{n,m}^{(i)}\}_{\substack{i=1,2,3;n\in\mathbb{N}_{0_i};\\ m=1,\ldots,2n+1}}$ we proceed in the same manner. The system of (Edmonds-)vector harmonics

$$u_{n,m}^{(i)} = (\nu_n^{(i)})^{-1/2} k_n^{(i)} Y_{n,m}, \tag{4.4.6}$$

$n \in \mathbb{N}_{0_i}$; $m = 1, \ldots, 2n + 1$; $i = 1, 2, 3$, forms an $l^2(\Omega)$–orthonormal system of vector spherical harmonics of kind i. Therefore we can write

$$\begin{aligned}
u^{(1)}_{n,m}(\xi) &= \sqrt{\mu_n}((n+1)(2n+1))^{-1/2}\, \mathrm{h}^{(1)}_{n,m}(x)|_{|x|=1} = \sqrt{\mu_n}((n+1)(2n+1))^{-1/2}\, \mathrm{h}^{(1)}_{n,m}(\xi),\\
u^{(2)}_{n,m}(\xi) &= \sqrt{\mu_n}(n(2n+1))^{-1/2}\, \mathrm{h}^{(2)}_{n,m}(x)|_{|x|=1} = \sqrt{\mu_n}(n(2n+1))^{-1/2}\, \mathrm{h}^{(2)}_{n,m}(\xi),\\
u^{(3)}_{n,m}(\xi) &= \sqrt{\mu_n}(n(n+1))^{-1/2}\, \mathrm{h}^{(3)}_{n,m}(x)|_{|x|=1} = \sqrt{\mu_n}(n(n+1))^{-1/2}\, \mathrm{h}^{(3)}_{n,m}(\xi),
\end{aligned}$$

where

$$\begin{aligned}
\mathrm{h}^{(1)}_{n,m}(x) &= ((2n+1)x - |x|^2\nabla_x)H_{n,m}(x),\\
\mathrm{h}^{(2)}_{n,m}(x) &= \nabla_x H_{n,m}(x),\\
\mathrm{h}^{(3)}_{n,m}(x) &= x \wedge \nabla_x H_{n,m}(x),
\end{aligned}$$

with $x = r\xi$, $r = |x|$, $\xi \in \Omega$, and $\nabla^* = lim_{r\to 1}(r\nabla_x - x\frac{\partial}{\partial r})$. In the same way as before our purpose is to determine the vector spherical harmonics with exact integer arithmetic. We start again with the representation

$$\mathrm{h}^{(i)}_{n,m}(x) = \sum_{k=1}^{3} \varepsilon^k \left(\sum_{[\alpha]=l_i} D^{i,k}_{\alpha;m} x^\alpha \right), \tag{4.4.7}$$

where we have used the abbreviations $l_1 = n+1$, $l_2 = n-1$, $l_3 = n$. For the vector harmonics we derive

$$u^{(1)}_{n,m}(\xi) = \sqrt{\frac{\mu_n}{\nu^{(1)}_n}} k^{(1)}_n H_{n,m}(r\xi)|_{r=1}, \tag{4.4.8}$$

$$u^{(2)}_{n,m}(\xi) = \sqrt{\frac{\mu_n}{\nu^{(2)}_n}} k^{(2)}_n H_{n,m}(r\xi)|_{r=1}, \tag{4.4.9}$$

$$u^{(3)}_{n,m}(\xi) = \sqrt{\frac{\mu_n}{\nu^{(3)}_n}} k^{(3)}_n H_{n,m}(r\xi)|_{r=1} \tag{4.4.10}$$

we find by observing the definition of the $k^{(i)}_n$-operators (see Definition 3.3.3)

$$\begin{aligned}
u^{(1)}_{n,m}(\xi) &= \sqrt{\frac{\mu_n}{\nu^{(1)}_n}} \sum_{[\alpha]=n} C^m_\alpha \begin{pmatrix} (2n+1)\xi_1^{\alpha_1+1}\xi_2^{\alpha_2}\xi_3^{\alpha_3} - \alpha_1\xi_1^{\alpha_1-1}\xi_2^{\alpha_2}\xi_3^{\alpha_3}(\xi_1^2+\xi_2^2+\xi_3^2)\\ (2n+1)\xi_1^{\alpha_1}\xi_2^{\alpha_2+1}\xi_3^{\alpha_3} - \alpha_2\xi_1^{\alpha_1}\xi_2^{\alpha_2-1}\xi_3^{\alpha_3}(\xi_1^2+\xi_2^2+\xi_3^2)\\ (2n+1)\xi_1^{\alpha_1}\xi_2^{\alpha_2}\xi_3^{\alpha_3+1} - \alpha_3\xi_1^{\alpha_1}\xi_2^{\alpha_2}\xi_3^{\alpha_3-1}(\xi_1^2+\xi_2^2+\xi_3^2) \end{pmatrix},\\
u^{(2)}_{n,m}(\xi) &= \sqrt{\frac{\mu_n}{\nu^{(2)}_n}} \sum_{[\alpha]=n} C^m_\alpha \begin{pmatrix} \alpha_1\xi_1^{\alpha_1-1}\xi_2^{\alpha_2}\xi_3^{\alpha_3}\\ \alpha_2\xi_1^{\alpha_1}\xi_2^{\alpha_2-1}\xi_3^{\alpha_3}\\ \alpha_3\xi_1^{\alpha_1}\xi_2^{\alpha_2}\xi_3^{\alpha_3-1} \end{pmatrix},\\
u^{(3)}_{n,m}(\xi) &= \sqrt{\frac{\mu_n}{\mu^{(3)}_n}} \sum_{[\alpha]=n} C^m_\alpha \begin{pmatrix} \alpha_3\xi_1^{\alpha_1}\xi_2^{\alpha_2+1}\xi_3^{\alpha_3-1} - \alpha_2\xi_1^{\alpha_1}\xi_2^{\alpha_2-1}\xi_3^{\alpha_3+1}\\ \alpha_1\xi_1^{\alpha_1-1}\xi_2^{\alpha_2}\xi_3^{\alpha_3+1} - \alpha_3\xi_1^{\alpha_1+1}\xi_2^{\alpha_2}\xi_3^{\alpha_3-1}\\ \alpha_2\xi_1^{\alpha_1+1}\xi_2^{\alpha_2-1}\xi_3^{\alpha_3} - \alpha_1\xi_1^{\alpha_1-1}\xi_2^{\alpha_1+1}\xi_3^{\alpha_3} \end{pmatrix}.
\end{aligned}$$

The determination of the coefficients $D^{i,k}_{\beta,m}$ follows immediately. We organize their computation again by a matrix-matrix multiplication:

$$D^{i,k}_{\beta;m} = \sqrt{\frac{\mu_n}{\mu_n^{(i)}}} \sum_{[\alpha]=n} C^m_\alpha M^{i,k}_{\beta\alpha;m}$$

for $i,k = 1,2,3$; $m = 1,\ldots,2n+1$,

$$[\beta] = \begin{cases} n+1, & for \quad i=1, \\ n-1, & for \quad i=2, \\ n, & for \quad i=3. \end{cases}$$

The matrices $\mathbf{m}^{i,k} = (M^{i,k}_{\beta,\alpha})$ have $\binom{n+2}{2}$ rows and $\binom{n}{2}$ columns in the cases $i=1$ and they have $\binom{n}{2}$ rows and $\binom{n}{2}$ columns for $i=2,3$. We get the following representation:

$$M^{1,1}_{\beta\alpha;m} = \begin{cases} (2n+1)-\alpha_1, & \text{if } \beta-\alpha = (1,0,0)^T, \\ -\alpha_1, & \text{if } \beta-\alpha \in \{(-1,2,0)^T, (-1,0,2)^T\}, \\ 0, & \text{otherwise,} \end{cases}$$

$$M^{1,2}_{\beta\alpha;m} = \begin{cases} (2n+1)-\alpha_2, & \text{if } \beta-\alpha = (0,1,0)^T, \\ -\alpha_2, & \text{if } \beta-\alpha \in \{(2,-1,0)^T, (0,-1,2)^T\}, \\ 0, & \text{otherwise,} \end{cases}$$

$$M^{1,3}_{\beta\alpha;m} = \begin{cases} (2n+1)-\alpha_3, & \text{if } \beta-\alpha = (0,0,1)^T, \\ -\alpha_3, & \text{if } \beta-\alpha \in \{(2,0,-1)^T, (0,2,-1)^T\}, \\ 0, & \text{otherwise,} \end{cases}$$

$$M^{2,1}_{\beta\alpha;m} = \begin{cases} \alpha_1, & \text{if } \beta-\alpha = (-1,0,0)^T, \\ 0, & \text{otherwise,} \end{cases}$$

$$M^{2,2}_{\beta\alpha;m} = \begin{cases} \alpha_2, & \text{if } \beta-\alpha = (0,-1,0)^T, \\ 0, & \text{otherwise,} \end{cases}$$

$$M^{2,3}_{\beta\alpha;m} = \begin{cases} \alpha_3, & \text{if } \beta-\alpha = (0,0,-1)^T, \\ 0, & \text{otherwise,} \end{cases}$$

$$M^{3,1}_{\beta\alpha;m} = \begin{cases} \alpha_3, & \text{if } \beta-\alpha = (0,1,-1)^T, \\ -\alpha_2, & \text{if } \beta-\alpha = (0,-1,1)^T, \\ 0, & \text{otherwise,} \end{cases}$$

$$M^{3,2}_{\beta\alpha;m} = \begin{cases} \alpha_1, & \text{if } \beta - \alpha = (-1,0,1)^T, \\ -\alpha_3, & \text{if } \beta - \alpha = (1,0,-1)^T, \\ 0, & \text{otherwise,} \end{cases}$$

$$M^{3,3}_{\beta\alpha;m} = \begin{cases} \alpha_2, & \text{if } \beta - \alpha = (1,-1,0)^T, \\ -\alpha_1, & \text{if } \beta - \alpha = (-1,1,0)^T, \\ 0, & \text{otherwise.} \end{cases}$$

We derive the vector harmonics from the scalar harmonics as given by Equation (4.3.44)-(4.3.50) and obtain

$$u^{(1)}_{3,1}(\xi) = \sqrt{\frac{105}{4\pi}}\frac{1}{\sqrt{24}}\begin{pmatrix} 7\xi_1{}^2(\xi_1{}^2 - 3\xi_2{}^2) - 3(\xi_1{}^2 - \xi_2{}^2) \\ 7\xi_1\xi_2(\xi_1{}^2 - 3\xi_2{}^2) + 6\xi_1\xi_2 \\ 7\xi_1\xi_3(\xi_1{}^2 - 3\xi_2{}^2) \end{pmatrix},$$

$$u^{(1)}_{3,2}(\xi) = \sqrt{\frac{105}{4\pi}}\begin{pmatrix} 7\xi_1{}^2\xi_2\xi_3 - \xi_2\xi_3 \\ 7\xi_1\xi_2{}^2\xi_3 - \xi_1\xi_3 \\ 7\xi_1\xi_2\xi_3{}^2 - \xi_1\xi_2 \end{pmatrix},$$

$$u^{(1)}_{3,3}(\xi) = \sqrt{\frac{105}{4\pi}}\frac{1}{\sqrt{40}}\begin{pmatrix} 7\xi_1{}^2(1 - 5\xi_3{}^2) - 3\xi_1{}^2 - \xi_2{}^2 + 4\xi_3{}^2 \\ 7\xi_1\xi_2(1 - 5\xi_3{}^2) - 2\xi_1\xi_2 \\ 7\xi_1\xi_3(1 - 5\xi_3{}^2) + 8\xi_1\xi_3 \end{pmatrix},$$

$$u^{(1)}_{3,4}(\xi) = \sqrt{\frac{105}{4\pi}}\frac{1}{\sqrt{24}}\begin{pmatrix} 7\xi_1\xi_2(3\xi_1{}^2 - \xi_2{}^2) - 6\xi_1\xi_2 \\ 7\xi_2{}^2(3\xi_1{}^2 - \xi_2{}^2) - 3\xi_1{}^2 + 3\xi_2{}^2 \\ 7\xi_2\xi_3(3\xi_1{}^2 + \xi_2{}^2) \end{pmatrix},$$

$$u^{(1)}_{3,5}(\xi) = \sqrt{\frac{105}{4\pi}}\frac{1}{\sqrt{4}}\begin{pmatrix} 7\xi_1\xi_3(\xi_1{}^2 - \xi_2{}^2) - 2\xi_1\xi_3 \\ 7\xi_2\xi_3(\xi_1{}^2 - \xi_2{}^2) + 2\xi_2\xi_3 \\ 7\xi_3(\xi_1{}^2 - \xi_2{}^2) - \xi_1{}^2 + \xi_2{}^2 \end{pmatrix},$$

$$u^{(1)}_{3,6}(\xi) = \sqrt{\frac{105}{4\pi}}\frac{1}{\sqrt{40}}\begin{pmatrix} 7\xi_1\xi_2(1 - 5\xi_3^2) - 2\xi_1\xi_2 \\ 7\xi_2{}^2(1 - 5\xi_3^2) - \xi_1{}^2 - 3\xi_2{}^2 + 4\xi_3{}^2 \\ 7\xi_2\xi_3(1 - 5\xi_3^2) + 8\xi_2\xi_3 \end{pmatrix},$$

$$u^{(1)}_{3,7}(\xi) = \sqrt{\frac{105}{4\pi}}\frac{1}{\sqrt{60}}\begin{pmatrix} 7\xi_1\xi_3(3 - 5\xi_3{}^2) - 6\xi_1\xi_3 \\ 7\xi_2\xi_3(3 - 5\xi_3{}^2) - 6\xi_2\xi_3 \\ 7\xi_3{}^2(3 - 5\xi_3{}^2) - 3(1 - 3\xi_3{}^2) \end{pmatrix},$$

and

$$
\begin{aligned}
u^{(2)}_{3,1}(\xi) &= \sqrt{\frac{105}{28\pi}}\frac{1}{\sqrt{24}}\begin{pmatrix} 3({\xi_1}^2-{\xi_2}^2) \\ 6\xi_1\xi_2 \\ 0 \end{pmatrix}, \\
u^{(2)}_{3,2}(\xi) &= \sqrt{\frac{105}{28\pi}}\begin{pmatrix} \xi_2\xi_3 \\ \xi_1\xi_3 \\ \xi_1\xi_2 \end{pmatrix}, \\
u^{(2)}_{3,3}(\xi) &= \sqrt{\frac{105}{28\pi}}\frac{1}{\sqrt{40}}\begin{pmatrix} 3\xi_1^2+\xi_2^2-4{\xi_3}^2 \\ 2\xi_1\xi_2 \\ -8\xi_1\xi_3 \end{pmatrix}, \\
u^{(2)}_{3,4}(\xi) &= \sqrt{\frac{105}{28\pi}}\frac{1}{\sqrt{24}}\begin{pmatrix} 6\xi_1\xi_2 \\ 3({\xi_1}^2-{\xi_2}^2) \\ 0 \end{pmatrix}, \\
u^{(2)}_{3,5}(\xi) &= \sqrt{\frac{105}{28\pi}}\frac{1}{\sqrt{4}}\begin{pmatrix} 2\xi_1\xi_3 \\ -2\xi_2\xi_3 \\ {\xi_1}^2-{\xi_2}^2 \end{pmatrix}, \\
u^{(2)}_{3,6}(\xi) &= \sqrt{\frac{105}{28\pi}}\frac{1}{\sqrt{40}}\begin{pmatrix} 2\xi_1\xi_2 \\ {\xi_1}^2+3{\xi_2}^2-4{\xi_3}^2 \\ -8\xi_2\xi_3 \end{pmatrix}, \\
u^{(2)}_{3,7}(\xi) &= \sqrt{\frac{105}{28\pi}}\frac{1}{\sqrt{60}}\begin{pmatrix} 6\xi_1\xi_3 \\ 6\xi_2\xi_3 \\ 3(1-3\xi_3^2) \end{pmatrix}.
\end{aligned}
$$

The vector spherical harmonics $u^{(i)}_{3,m}$ for $i = 1,2$; $m = 1\ldots,2n+1$, are illustrated in Figures 4.4.6- 4.4.7. Figures 4.4.8- 4.4.9 show vector spherical harmonics $u^{(i)}_{10,15}$ for $i = 1,2$. The functions of type 3 are omitted because they are equivalent to $y^{(3)}_{10,15}$ as shown in Figure 4.4.5. Each field $u^{(i)}_{n,m}$, $i = 1,2$, is a composition of vector spherical harmonics $y^{(i)}_{n,m}$. Therefore the fields $u^{(i)}_{n,m}$ of type $i = 1$ and $i = 2$ have radial and tangential contributions. Thus, we do not have any more the separation like in the case of $y^{(i)}_{n,m}$, $i = 1,2,3$ into a radial and tangential part. But still $u^{(3)}_{n,m}$ is a tangential field. In Figures 4.4.10-4.4.13 these properties are illustrated.

With the generated set of vector spherical harmonics we can proceed to extend them to

arbitrary spheres and finally obtain vector outer harmonics.
The extension from a sphere with $R_1 = 1$ to a sphere with $R_2 = 1.3$ is shown for $u^{(3)}_{5,6}$, $u^{(1)}_{5,9}$ and $u^{(3)}_{5,9}$ in Figure 4.4.14 .
To get the vector outer harmonics we see with (4.2.5) that we have to distinguish the factors for the different types $i = 1, 2, 3$, where the factor is given by $\left(\frac{R}{|x|}\right)^{n-2+l_i}$ with $l_i = i + 3(1 - 0_i)$, i.e.,

$$h^{(i),R}_{n,m}(x) = \frac{1}{R}\left(\frac{R}{|x|}\right)^{n-2+l_i} u^{(i)}_{n,m}\left(\frac{x}{|x|}\right), \qquad (4.4.11)$$

$n \in \mathbb{N}_{0_i}$; $m = 1, \ldots, 2n + 1$; $i = 1, 2, 3$.
Figures 4.4.15 and 4.4.16 show examples for $h^{(i),R}_{5,3}$, $i = 1, 3$ and $h^{(i),R}_{5,11}$, $i = 1, 3$, respectively.

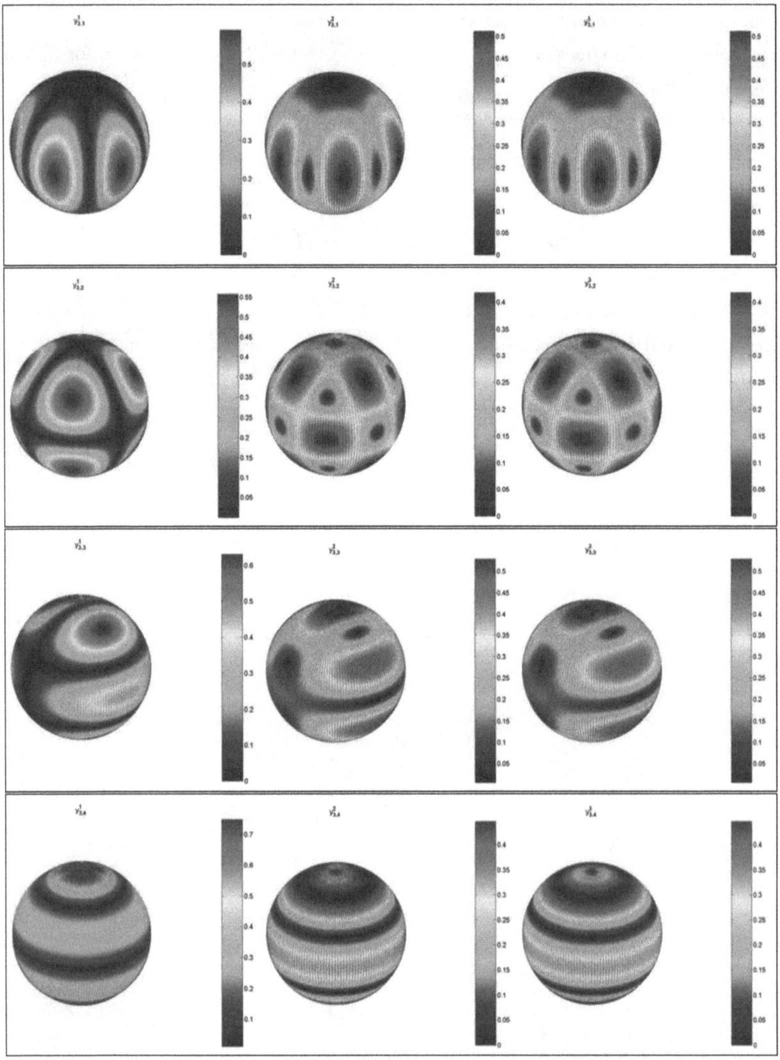

Figure 4.4.1: Vector spherical harmonics $y^{(i)}_{3,m}$ for $i = 1, 2, 3$, $m = 1, 2, 3, 4$. The columns represent the type $i = 1, 2, 3$ and the rows the order $m = 1, 2, 3, 4$. The colorbar represents the absolute value of the vectors.

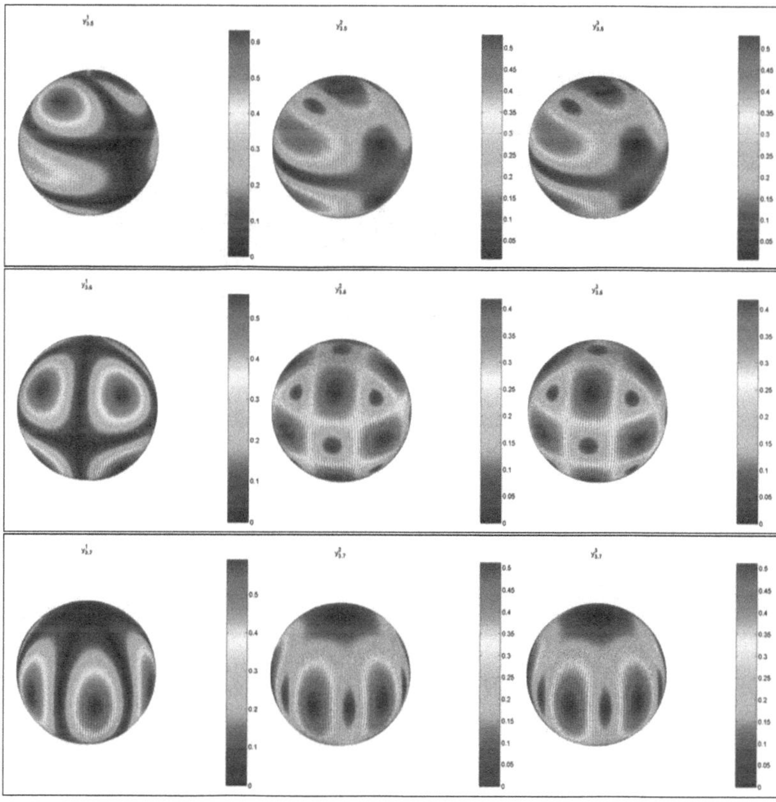

Figure 4.4.2: Vector spherical harmonics $y^{(i)}_{3,m}$ for $i = 1, 2, 3$, $m = 5, 6, 7$. The columns represent the type $i = 1, 2, 3$ and the rows the order $m = 5, 6, 7$. The colorbar represents the absolute value of the vectors.

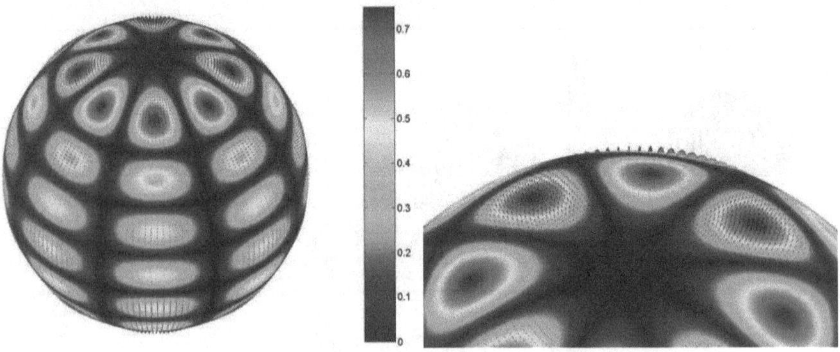

Figure 4.4.3: Vector spherical harmonic $y_{10,15}^{(1)}$ on the unit sphere Ω. At the right a close-up view around the pole area is provided.

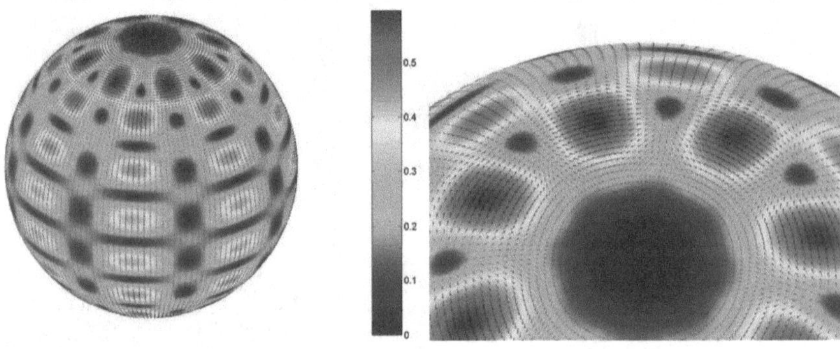

Figure 4.4.4: Vector spherical harmonic $y_{10,15}^{(2)}$ on the unit sphere Ω. At the right a close-up view around the pole area is provided.

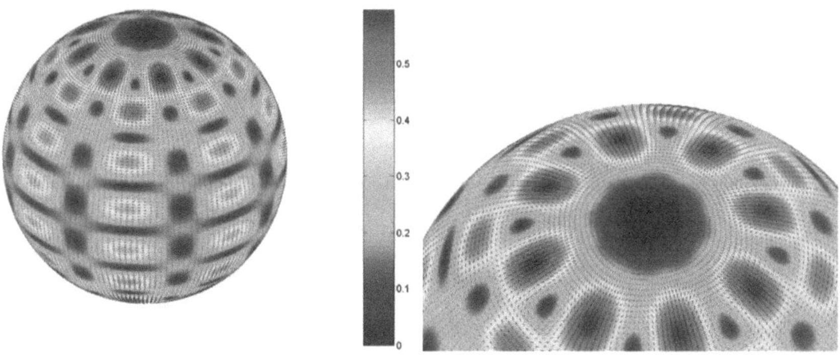

Figure 4.4.5: Vector spherical harmonic $y^{(3)}_{10,15}$ on the unit sphere Ω. At the right a close-up view around the pole area is provided.

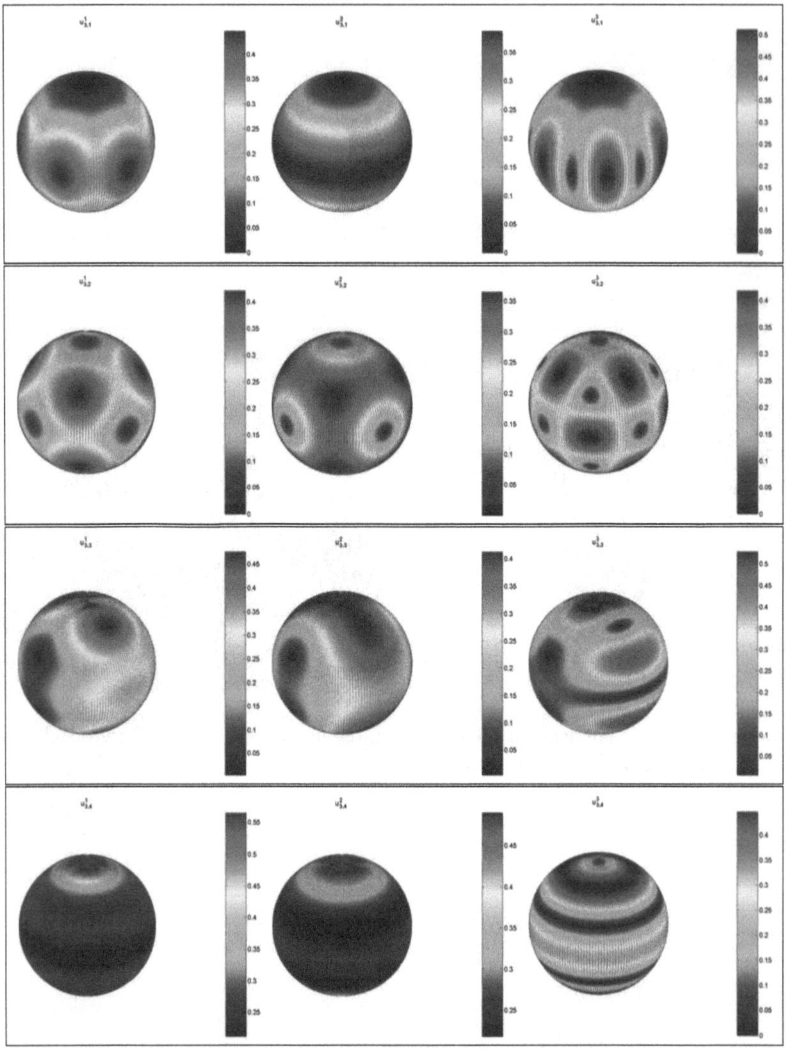

Figure 4.4.6: Vector spherical harmonics $u^{(i)}_{3,m}$ for $i = 1, 2, 3$, $m = 1, 2, 3, 4$. The columns represent the type $i = 1, 2, 3$ and the rows the order $m = 1, 2, 3, 4$. The colorbar represents the absolute value of the vectors.

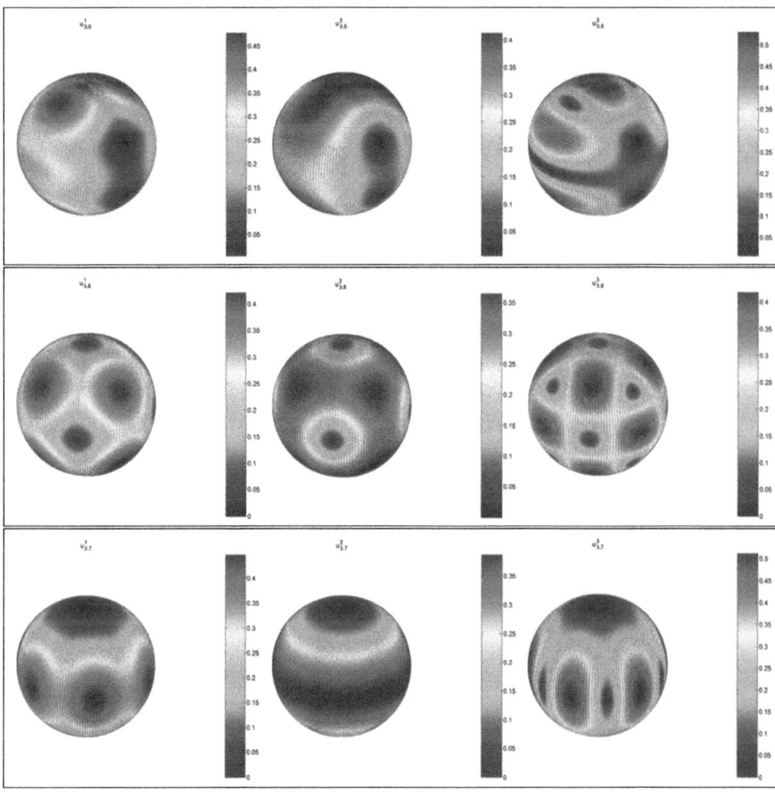

Figure 4.4.7: Vector spherical harmonics $u_{3,m}^{(i)}$ for $i = 1, 2, 3$, $m = 5, 6, 7$. The columns represent the type $i = 1, 2, 3$ and the rows the order $m = 5, 6, 7$. The colorbar represents the absolute value of the vectors.

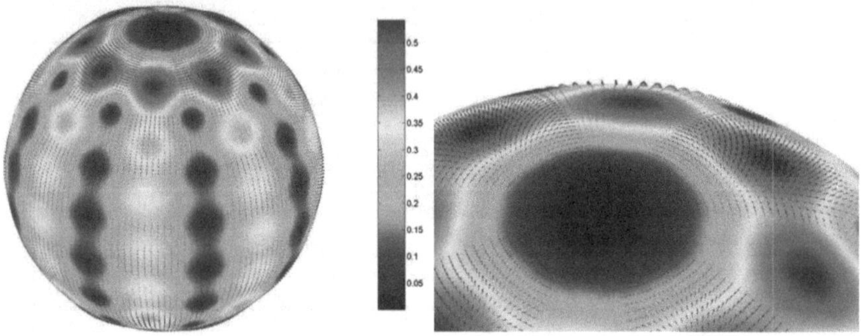

Figure 4.4.8: Vector spherical harmonic $u^{(1)}_{10,15}$ on the unit sphere Ω. At the right a close-up view around the pole area is provided.

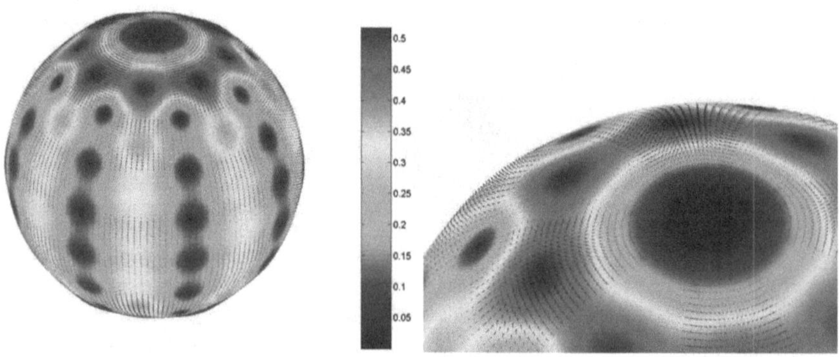

Figure 4.4.9: Vector spherical harmonic $u^{(2)}_{10,15}$ on the unit sphere Ω. At the right a close-up view around the pole area is provided.

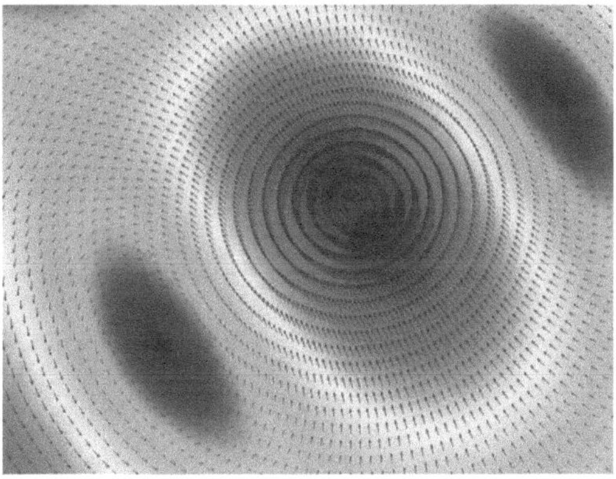

Figure 4.4.10: Zoomed section on the polar region of the unit sphere Ω with calculated values of $u^{(3)}_{5,1}$. The arrow inside the smallest circle (pole) illustrates the non singularity of the calculated vector spherical harmonic.

Figure 4.4.11: Zoomed section on the polar region of the unit sphere Ω with calculated values of $u^{(1)}_{29,5}$. The absolute value of the vectors is colored and ranges between 0 and 0.8202. Because it is of type one the direction of the vector field is neither pure tangential nor normal.

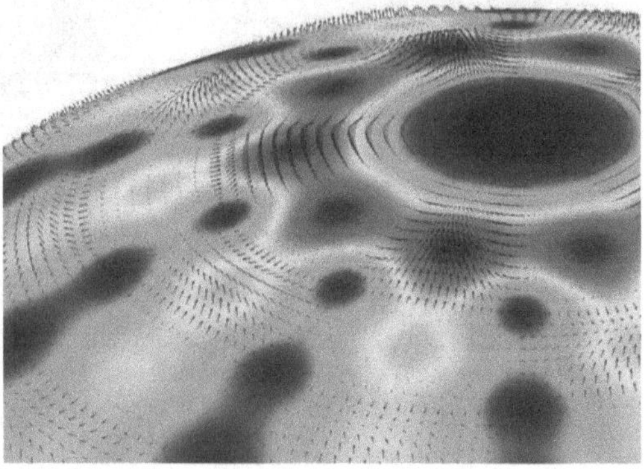

Figure 4.4.12: Zoomed part of a vector spherical harmonic $u^{(2)}_{29,5}$ on the unit sphere Ω. The absolute value of the vectors is colored and ranges between 0 and 0.8064. Because it is of type two the direction of the vector field is neither pure tangential nor normal.

Figure 4.4.13: Zoomed section on the polar region of the unit sphere Ω with calculated values of $u^{(3)}_{29,5}$. The absolute value of the vectors is colored and ranges between 0 and 0.751. Because it is of type three the direction of the vector field is pure tangential.

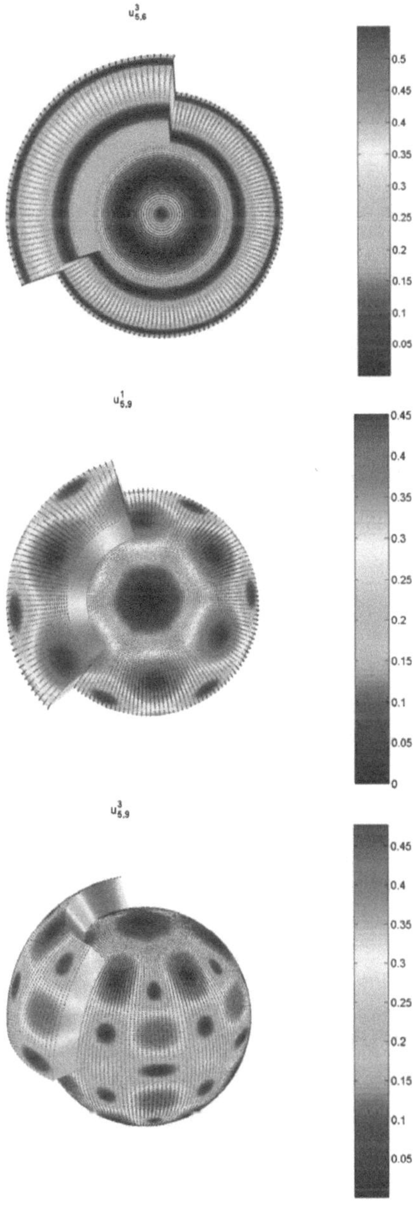

Figure 4.4.14: Illustration of the partly extension of vector spherical harmonics $u_{5,6}^{(3),R}$, $u_{5,9}^{(1),R}$ and $u_{5,9}^{(3),R}$ (from top to bottom) from a sphere $R_1 = 1$ to a sphere with $R_2 = 1.3$.

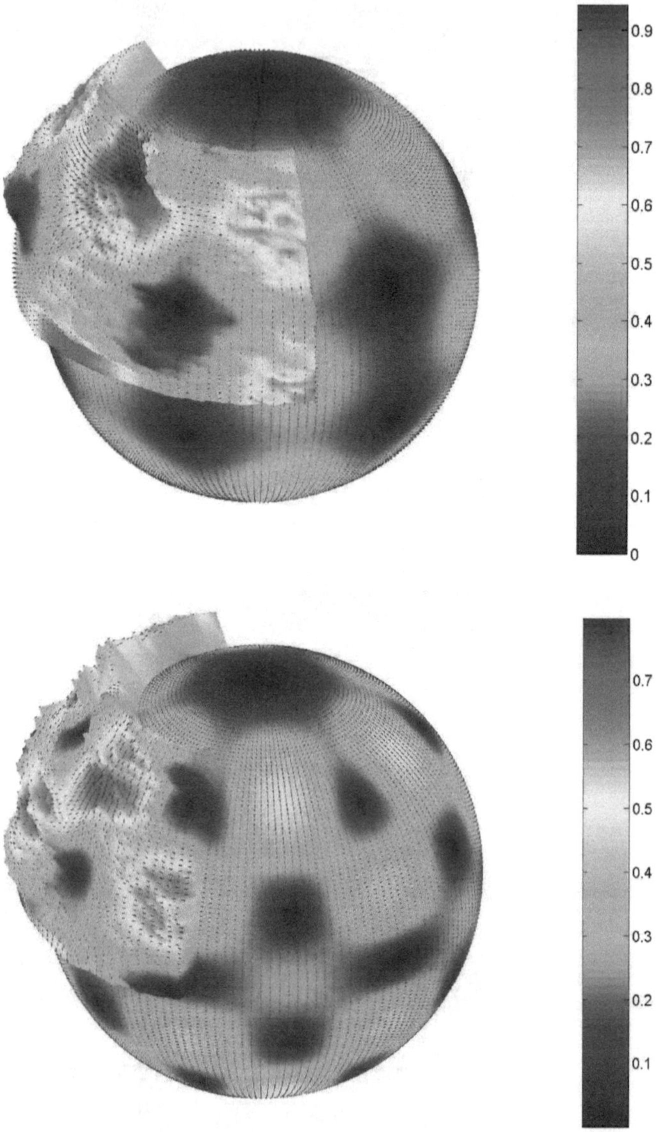

Figure 4.4.15: Illustration of vector outer harmonics $h_{5,3}^{(i),R}$, $i = 1, 3$, by replacing a part of the unit sphere by a regular surface (from top to bottom).

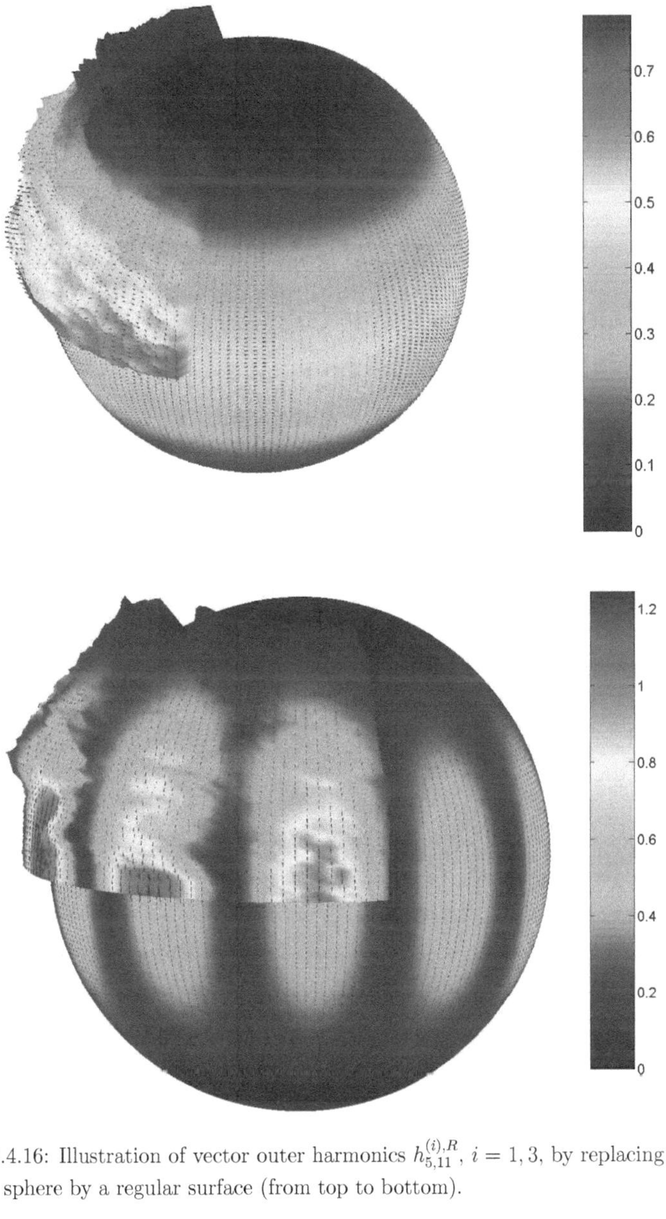

Figure 4.4.16: Illustration of vector outer harmonics $h_{5,11}^{(i),R}$, $i = 1, 3$, by replacing a part of the unit sphere by a regular surface (from top to bottom).

Chapter 5

Approximation of Vector Functions on Regular Surfaces

In this chapter, first, a Fourier representation of vector outer harmonics will be used for the approximation of continuous vector fields on regular surfaces (e.g., geoscientifically relevant surfaces like ellipsoid, geoid, real Earth's surface). This concept is characterized by the fact that an integral over a sphere, inside the regular surface, will be expressed by approximate formulae involving data points on a regular surface Σ (i.e., not on the sphere Ω_R). In other words our approach to approximate continuous vector fields on regular surfaces in terms of vector outer harmonics leads to a new class of approximate integration formulae, where the nodal points are not taken on the reference area of integration. This, of course, requires the solution of a linear system in terms of vector outer harmonics relating the integral over a sphere Ω_R, with $\Omega_R^{int} \subset \Sigma^{int}$ and $dist(\Omega_R, \Sigma) > 0$, i.e., the so called Runge sphere, to a (cubature) formula on a regular surface Σ. Second, the Fourier approach leads us to a minimum interpolation procedure of vector fields on a regular surface Σ based on a finite set of discretely given vector data on Σ. Essential tool is the reproducing kernel structure in the reference space h. The consideration closely parallels the interpolation method proposed by [13]. Referring to Figure 1.0.1 we deal in this chapter with step (4) and (5).

We begin our considerations with preparatory material such as the reference space h for vector fields harmonic in the outer space of the Runge sphere Ω_R. Similar approaches for scalar fields can be found, e.g., in [18], [21], and [29].

5.1 Reproducing Kernel Structure of the Reference Space h

In what follows we introduce the reference space h in which the approximation of vector fields on regular surfaces will be done (see, e.g., approach [13]).
Let f be a field of class $l^2(\Omega)$. Suppose that Σ is a regular surface satisfying (2.0.33). Then it can easily be seen that the sum

$$v(x) = \sum_{i=1}^{3} \sum_{n=0_i}^{\infty} \sum_{m=1}^{2n+1} (f, h_{n,m}^{(i),R})_{l^2(\Omega)} h_{n,m}^{(i),R}(x)$$

is absolutely and uniformly convergent for all $x \in \overline{\Sigma_{inf}^{ext}}$. The linear space h defined by

$$h = \{v \,|\, v = \sum_{i=1}^{3} \sum_{n=0_i}^{\infty} \sum_{m=1}^{2n+1} (f, h_{n,m}^{(i),R})_{l^2(\Omega)} h_{n,m}^{(i),R}(x)|_{\overline{\Sigma_{inf}^{ext}}} \, , \; f \in l^2(\Omega)\} \tag{5.1.1}$$

is a separable Hilbert space with respect to the inner space $(\cdot,\cdot)_h$ corresponding to the norm

$$\begin{aligned} \|v\|_h = \|f\|_{l^2(\Omega_R)} &= \left(\int_{\Omega_R} \left| \sum_{i=1}^{3} (f(x) \cdot \varepsilon^i) \right|^2 d\omega(x) \right)^{1/2} \\ &= \left(\int_{\Omega_R} f(x) \cdot f(x) \, d\omega(x) \right)^{1/2} . \end{aligned} \tag{5.1.2}$$

Theorem 5.1.1. *The space h defined by (5.1.1) with the norm (5.1.2) is a Hilbert space. The kernel $\mathbf{k}(\cdot,\cdot)$ given by*

$$\mathbf{k}(x,y) = \sum_{i=1}^{3} \sum_{n=0_i}^{\infty} \sum_{m=1}^{2n+1} h_{n,m}^{(i),R}(x) \otimes h_{n,m}^{(i),R}(y), \quad x, y \in \overline{\Sigma_{inf}^{ext}}, \tag{5.1.3}$$

is a reproducing kernel for h in the sense that

(i) for fixed $x \in \overline{\Sigma_{inf}^{ext}}$ each vector field $\mathbf{k}(\cdot,x)\varepsilon^i$, $i = 1,2,3$, given by

$$\mathbf{k}(\cdot,x)\varepsilon^i|_{\overline{\Sigma_{inf}^{ext}}} = \sum_{i=1}^{3} \sum_{n=0_i}^{\infty} \sum_{m=1}^{2n+1} \left(h_{n,m}^{(i),R}(x) \cdot \varepsilon^i\right) h_{n,m}^{(i),R}|_{\overline{\Sigma_{inf}^{ext}}}$$

is an element of h.

(ii) for every $v \in h$ and every $x \in \overline{\Sigma_{inf}^{ext}}$, the reproducing property

$$v(x) = \sum_{i=1}^{3} \left(\mathbf{k}(\cdot, x)\varepsilon^i, v\right)_h \varepsilon^i$$

is valid, so that

$$v(x)\varepsilon^i = \left(\mathbf{k}(\cdot, x)\varepsilon^i, v\right)_h, \quad i = 1, 2, 3. \tag{5.1.4}$$

Remark 5.1.1. *By application of the Cauchy-Schwarz inequality, we obtain*

$$|v(x)\varepsilon^i| \leq \left(\varepsilon^i \cdot \mathbf{k}(x, x)\varepsilon^i\right)^{1/2} \|v\|_h.$$

This means that

$$|v(x)| \leq \left(\sum_{i=1}^{3} \varepsilon^i \cdot \mathbf{k}(x, x)\varepsilon^i\right)^{1/2} \|v\|_h$$

for every $x \in \overline{\Sigma_{inf}^{ext}}$ and all $v \in h$. Thus, we are able to deduce that there exists a positive constant A (dependent on Σ, but independent of v) such that

$$\sup_{x \in \overline{\Sigma^{ext}}} |v(x)| \leq A\|v\|_h.$$

Consider the linear space spanned by the vector fields $h_{n,m}^{(i),R}$, $n \in \mathbb{N}_{0_i}$; $m = 1, \ldots, 2n+1$; $i = 1, 2, 3$,

$$h_{0,\ldots,a} = \{v \,|\, v = \sum_{i=1}^{3} \sum_{n=0_i}^{a_i} \sum_{m=1}^{2n+1} (f, h_{n,m}^{(i),R})_{l^2(\Omega_R)} \, h_{n,m}^{(i),R}, \ f \in l^2(\Omega_R)\}$$

with a_i being defined in connection with Lemma 3.3.12 by

$$a_i = \begin{cases} a-1, & \text{if } i = 1, \\ a+1, & \text{if } i = 2, \\ a, & \text{if } i = 3 \end{cases} \tag{5.1.5}$$

for $a > 1$.

Of course, $h_{0,\ldots,a}$ possesses the dimension

$$M = 3(a+1)^2, \ a > 1.$$

Moreover, $h_{0,\dots,a}$, $a>1$, is a M-dimensional Hilbert space equipped with the inner product $(\cdot,\cdot)_{h_{0,\dots,a}}$, given by

$$(v_1,v_2)_{h_{0,\dots,a}} = \sum_{i=1}^{3}\sum_{n=0_i}^{a_i}\sum_{m=1}^{2n+1} (f_1, h_{n,m}^{(i),R})_{l^2(\Omega_R)} (f_2, h_{n,m}^{(i),R})_{l^2(\Omega_R)}, \quad f_1,f_2\in l^2(\Omega_R),$$

and the reproducing kernel

$$\mathbf{k}_{h_{0,\dots,a}}(x,y) = \sum_{i=1}^{3}\sum_{n=0_i}^{a_i}\sum_{m=1}^{2n+1} h_{n,m}^{(i),R}(x)\otimes h_{n,m}^{(i),R}(y), \quad x,y\in\overline{\Sigma_{inf}^{ext}}. \tag{5.1.6}$$

The tensor kernels (5.1.6) are related to the (Morse-Feshbach)-vector spherical harmonics $y_{n,m}^{(i)}$, $n\in\mathbb{N}_{0_i}$, $m=1,\dots,2n+1$, $i=1,2,3$, in the following way:

Corollary 5.1.2. *Let $y_{n,m}^{(i)}$, $n\in\mathbb{N}_{0_i}$, $m=1,\dots,2n+1$, $i=1,2,3$, $a>1$, be vector spherical harmonics as given by Theorem 3.3.1 then*

$$\begin{aligned}
&\sum_{i=1}^{3}\mathbf{k}^{(i)}_{h_{0,\dots,a}}(x,y) = \\
&\quad \frac{R}{|x||y|}\frac{1}{3}\Big(y_{n,m}^{(1)}(\xi)\otimes y_{n,m}^{(1)}(\eta) + \sqrt{2}(y_{n,m}^{(1)}(\xi)\otimes y_{n,m}^{(2)}(\eta) + y_{n,m}^{(2)}(\xi)\otimes y_{n,m}^{(1)}(\eta)) + \\
&\qquad\qquad 2y_{n,m}^{(3)}(\xi)\otimes y_{n,m}^{(3)}(\eta)\Big) + \\
&\quad \sum_{i=1}^{3}\sum_{n=1}^{a}\sum_{m=1}^{2n+1} R_n^{(i)} y_{n,m}^{(i)}(\xi)\otimes y_{n,m}^{(i)}(\eta) + R_n^{(1,2)}\left(y_{n,m}^{(1)}(\xi)\otimes y_{n,m}^{(2)}(\eta) + y_{n,m}^{(2)}(\xi)\otimes y_{n,m}^{(1)}(\eta)\right).
\end{aligned}$$

where

$$\begin{aligned}
R_n^{(1)} &= \left(\frac{R}{|x||y|}\right)^{n+1}\left(\frac{(2n+1)^2-2}{(2n+1)^2-4}\right),\\
R_n^{(2)} &= \left(\frac{R}{|x||y|}\right)^{n+1}\left(\frac{(2n+1)^2-6}{(2n+1)^2-4}\right),\\
R_n^{(3)} &= \left(\frac{R}{|x||y|}\right)^{n+1},\\
R_n^{(1,2)} &= \left(\frac{R}{|x||y|}\right)^{n+1}\left(\frac{\sqrt{(2n+1)^2(n+1)(n+2)}-\sqrt{(2n+3)^2 n(n-1)}}{(2n+1)^2-4}\right).
\end{aligned}$$

Proof. We calculate the kernel for each pair of i, $i=1,2,3$, using the representation of outer harmonics in terms of (Edmonds-)vector spherical harmonics (see (4.2.5)) and by substituting these vectors by the (Morse-Feshbach)-vector spherical harmonics as given by (3.3.20). We set $x = |x|\xi$ and $y = |y|\eta$, $\xi, \eta \in \Omega$ obtaining the following expressions:

$$\begin{aligned}
\mathbf{k}^{(1)}_{h_{0,\ldots,a}}(x,y) &= \sum_{n=0}^{a_1}\sum_{m=1}^{2n+1} h^{(1)}_{n,m}(x) \otimes h^{(1)}_{n,m}(y) \\
&= \sum_{n=1}^{a}\sum_{m=1}^{2n+1} \left(\frac{R}{|x||y|}\right)^{n+1} \Bigg[\frac{n}{2n-1} y^{(1)}_{n,m}(\xi) \otimes y^{(1)}_{n,m}(\eta) - \\
&\quad \frac{\sqrt{n(n-1)}}{2n-1}\left(y^{(1)}_{n,m}(\xi) \otimes y^{(2)}_{n,m}(\eta) + y^{(2)}_{n,m}(\xi) \otimes y^{(1)}_{n,m}(\eta)\right) + \\
&\quad \frac{n-1}{2n-1} y^{(2)}_{n,m}(\xi) \otimes y^{(2)}_{n,m}(\eta)\Bigg],
\end{aligned}$$

$$\begin{aligned}
\mathbf{k}^{(2)}_{h_{0,\ldots,a}}(x,y) &= \sum_{n=1}^{a_2}\sum_{m=1}^{2n+1} h^{(2)}_{n,m}(x) \otimes h^{(2)}_{n,m}(y) \\
&= \sum_{n=0}^{a}\sum_{m=1}^{2n+1} \left(\frac{R}{|x||y|}\right)^{n+1} \Bigg[\frac{n+1}{2n+3} y^{(1)}_{n,m}(\xi) \otimes y^{(1)}_{n,m}(\eta) + \\
&\quad \frac{\sqrt{(n+1)(n+2)}}{2n+3}\left(y^{(1)}_{n,m}(\xi) \otimes y^{(2)}_{n,m}(\eta) + y^{(2)}_{n,m}(\xi) \otimes y^{(1)}_{n,m}(\eta)\right) + \\
&\quad \frac{n+2}{2n+3} y^{(2)}_{n,m}(\xi) \otimes y^{(2)}_{n,m}(\eta)\Bigg],
\end{aligned}$$

$$\mathbf{k}^{(3)}_{h_{0,\ldots,a}}(x,y) = \sum_{n=1}^{a_3}\sum_{m=1}^{2n+1} h^{(3)}_{n,m}(x) \otimes h^{(3)}_{n,m}(y) = \sum_{n=1}^{a}\sum_{m=1}^{2n+1} \left(\frac{R}{|x||y|}\right)^{n+1} y^{(3)}_{n,m}(\xi) \otimes y^{(3)}_{n,m}(\eta).$$

Building the sum with the above presented kernels and splitting the sum for the type $i = 2$ into $n = 0$ and $n = 1, \ldots, a$, $a > 1$, we obtain:

$$\begin{aligned}
\sum_{i=1}^{3} \mathbf{k}^{(i)}_{h_{0,\ldots,a}}(x,y) &= \mathbf{k}^{(1)}_{h_{0,\ldots,a}}(x,y) + \mathbf{k}^{(2)}_{h_{0,\ldots,a}}(x,y) + \mathbf{k}^{(3)}_{h_{0,\ldots,a}}(x,y) \\
&= \frac{R}{|x||y|}\frac{1}{3}\Big(y^{(1)}_{n,m}(\xi) \otimes y^{(1)}_{n,m}(\eta) + \sqrt{2}(y^{(1)}_{n,m}(\xi) \otimes y^{(2)}_{n,m}(\eta) + y^{(2)}_{n,m}(\xi) \otimes y^{(1)}_{n,m}(\eta)) + \\
&\quad 2y^{(3)}_{n,m}(\xi) \otimes y^{(3)}_{n,m}(\eta)\Big) +
\end{aligned}$$

$$\sum_{n=1}^{a}\sum_{m=1}^{2n+1}\left(\frac{R}{|x||y|}\right)^{n+1}\left[\left(\frac{(2n+1)^2-2}{(2n+1)^2-4}\right)y^{(1)}_{n,m}(\xi)\otimes y^{(1)}_{n,m}(\eta)+\right.$$
$$\left(\frac{(2n+1)^2-6}{(2n+1)^2-4}\right)y^{(2)}_{n,m}(\xi)\otimes y^{(2)}_{n,m}(\eta)+y^{(3)}_{n,m}(\xi)\otimes y^{(3)}_{n,m}(\eta)+$$
$$\left.\left(\frac{\sqrt{(2n+1)^2(n+1)(n+2)}-\sqrt{(2n+3)^2n(n-1)}}{(2n+1)^2-4}\right)\left(y^{(1)}_{n,m}(\xi)\otimes y^{(2)}_{n,m}(\eta)+y^{(2)}_{n,m}(\xi)\otimes y^{(1)}_{n,m}(\eta)\right)\right].$$

□

Denote by $h^{\perp}_{0,\dots,a}$ the orthogonal compliment of $h_{0,\dots,a}$ in h. The linear space with inner product $(\cdot,\cdot)_{h^{\perp}_{0,\dots,a}}$ defined by

$$(v_1,v_2)_{h^{\perp}_{0,\dots,a}}=\sum_{i=1}^{3}\sum_{n=a_i+1}^{\infty}\sum_{m=1}^{2n+1}(f_1,h^{(i),R}_{n,m})_{l^2(\Omega_R)}(f_2,h^{(i),R}_{n,m})_{l^2(\Omega_R)},\quad f_1,f_2\in l^2(\Omega_R),$$

and reproducing kernel

$$\mathbf{k}_{h^{\perp}_{0,\dots,a}}(x,y)=\sum_{i=1}^{3}\sum_{n=a_i+1}^{\infty}\sum_{m=1}^{2n+1}h^{(i),R}_{n,m}(x)\otimes h^{(i),R}_{n,m}(y),\quad x,y\in\overline{\Sigma^{ext}_{inf}}.$$

Hence, h is the orthogonal direct sum of $h_{0,\dots,a}$ and $h^{\perp}_{0,\dots,a}$

$$h=h_{0,\dots,a}\otimes h^{\perp}_{0,\dots,a}$$

with inner product

$$(v_1,v_2)_h=(v_1,v_2)_{h_{0,\dots,a}}+(v_1,v_2)_{h^{\perp}_{0,\dots,a}}$$

and the reproducing kernel

$$\mathbf{k}(x,y)=\mathbf{k}_{h_{0,\dots,a}}(x,y)+\mathbf{k}_{h^{\perp}_{0,\dots,a}}(x,y).\qquad(5.1.7)$$

For the representation of the kernel $\mathbf{k}_{h^{\perp}_{0,\dots,a}}(x,y)$ considerations as proposed in [2] can be taken into account.

5.2 Fourier Representation of Vector Functions on Regular Surfaces

In this chapter we are interested in Fourier representations of vector functions on regular surfaces. Our work is based on the scalar case in [18, 21]. We already mentioned that the set of all finite linear combinations of vector outer harmonics restricted to a regular surface Σ is uniformly dense in the space $c^{(0)}(\Sigma)$. What follows is that the refernce space $h|_\Sigma$ is also a uniformly dense subset of $c^{(0)}(\Sigma)$.

Suppose that from a continuous function v on a regular surface Σ a discrete set of function values $v(x_1) = v_1, \ldots, v(x_N) = v_N$, $v_i \in \mathbb{R}^3$, $i = 1, \ldots, N$, are available on the set of points $\{x_1, \ldots, x_N\} \subset \Sigma$. We are interested in the approximation of a continuous vector function on a regular surface Σ corresponding to the scattered vector function values on the finite set of discrete points on Σ.
The closure and completeness of vector outer harmonics in connection with Helly's theorem [37] shows that, corresponding to the continuous vector function v on Σ, there exists a member u of class $h|_\Sigma$ in an $(\varepsilon/2)-$neighborhood, such that the values of u are consistent with the function values of the continuous vector function v on Σ for the known finite set of discrete points, i.e., $u(x_i) = v_i = v(x_i)$, $i = 1, \ldots, N$. Moreover, this function u of class $h|_\Sigma$ may be considered to be in $(\varepsilon/2)-$accuracy to a member $u_{0,\ldots,a}$ of class $h_{0,\ldots,a}|_\Sigma$ which can be supposed to be consistent with the known function values as well, i.e., $u_{0,\ldots,a}(x_i) = v_i = v(x_i)$, $i = 1, \ldots, N$. Thus, to any continuous vector function v on a regular surface Σ, there exists in ε-accuracy a bandlimited vector function $u_{0,\ldots,a} \in h_{0,\ldots,a}|_\Sigma$ such that this bandlimited vector function coincides at all given points with the function values of the original continuous vector function on the regular surface Σ.
Our interest, first, lies in a Fourier approximation of a function $u_{0,\ldots,a}$ of class $h_{0,\ldots,a}|_\Sigma$ from discretely given vector function values on Σ. The method presented here is a generalization of the scalar Fourier variant (second variant of the paper [21]) due to Freeden and Schneider.

We start with the discussion of a new class of approximate formulae involving vector outer harmonics. For that purpose consider a vector function $v_{0,\ldots,a}$ of class $h_{0,\ldots,a}$ and a vector function u of class h. Our purpose is to develop a rule of the form

$$(v_{0,\ldots,a}, u)_h = \sum_{k=1}^{N} a_k \cdot v_{0,\ldots,a}(x_k), \tag{5.2.1}$$

where the weights $a_k \in \mathbb{R}^3$, $k = 1, \ldots, N$ are vectors with $3N \geq M$, $M = 3(a+1)^2$, $a > 1$ and $x_1, \ldots, x_N$ are knots on the regular surface Σ.

Let us define an admissible system for a set of vector outer harmonics. (Note that $(v_{0,\ldots,a}, u)_h$ can be written as an integral over Ω_R, hence, we are interested in integration rules of Ω_R corresponding to knots on the regular surface Σ).

Definition 5.2.1. *A system $\{x_1, \ldots, x_N\} \subset \Sigma$ of N points with $3N \geq M$, $M = 3(a+1)^2$, $a > 1$, is called an admissible system of order a on Σ with respect to $h_{0,\ldots,a}$, if the rank of the $(M, 3N)$-matrix*

$$\boldsymbol{h} = \begin{pmatrix} (h_{0,1}^{(1),R}(x_1))^T & \ldots & (h_{0,1}^{(1),R}(x_N))^T \\ \vdots & & \vdots \\ (h_{a_1,2a_1+1}^{(1),R}(x_1))^T & \ldots & (h_{a_1,2a_1+1}^{(1),R}(x_N))^T \\ (h_{1,1}^{(2),R}(x_1))^T & \ldots & (h_{1,1}^{(2),R}(x_N))^T \\ \vdots & & \vdots \\ (h_{a_2,2a_2+1}^{(2),R}(x_1))^T & \ldots & (h_{a_2,2a_2+1}^{(2),R}(x_N))^T \\ (h_{1,1}^{(3),R}(x_1))^T & \ldots & (h_{1,1}^{(3),R}(x_N))^T \\ \vdots & & \vdots \\ (h_{a_3,2a_3+1}^{(3),R}(x_1))^T & \ldots & (h_{a_3,2a_3+1}^{(3),R}(x_N))^T \end{pmatrix} \tag{5.2.2}$$

is equal to M, where $h_{n,m}^{(i)}$, $n \in \mathbb{N}_{0_i}$; $m = 1, \ldots, 2n+1$; $i = 1, 2, 3$, are the vector outer harmonics.

Definition (5.2.1) leads us to the following integration formula.

Lemma 5.2.1. *Let $\{x_1, \ldots, x_N\} \subset \Sigma$, $3N \geq M$, $M = 3(a+1)^2$, $a > 1$, be an admissible system of order a on Σ with respect to $h_{0,\ldots,a}$. Furthermore, suppose that $v_{0,\ldots,a} \in h_{0,\ldots,a}$ and $u \in h$. Then,*

$$(v_{0,\ldots,a}, u)_h = \sum_{j=1}^{3} \sum_{p=0_j}^{a_j} \sum_{q=1}^{2p+1} u^{(j)\wedge}(p,q) \sum_{k=1}^{N} a_k^{j,p,q} \cdot v_{0,\ldots,a}(x_k)$$

holds for all weights $a_1^{j,p,q}, \ldots, a_N^{j,p,q}$; $p = 0_j, \ldots, a_j$; $q = 1, \ldots, 2p+1$; $j = 1, 2, 3$, satisfying the linear equations

$$\sum_{k=1}^{N} a_k^{j,p,q} \cdot h_{n,m}^{(i)}(R; x_k) = \delta_{ij} \delta_{np} \delta_{mq} \tag{5.2.3}$$

for $n = 0_i, \ldots, a_i$*;* $m = 1, \ldots, 2n+1$*;* $i = 1, 2, 3$*.*

Proof. Using Fourier expansion of $v_{0,\ldots,a}$ and u together with the Parseval identity we are able to deduce that

$$(v_{0,\ldots,a}, u)_h = \sum_{i=1}^{3} \sum_{n=0_i}^{a_i} \sum_{m=1}^{2n+1} (v_{0,\ldots,a})^{(i)\wedge}(n,m) u^{(i)\wedge}(n,m),$$

where

$$(v_{0,\ldots,a})^{(i)\wedge}(n,m) = (v_{0,\ldots,a}, h_{n,m}^{(i),R})_h,$$
$$u^{(i)\wedge}(n,m) = (u, h_{n,m}^{(i),R})_h.$$

Therefore,

$$\begin{aligned}
(v_{0,\ldots,a}, u)_h &= \sum_{i=1}^{3} \sum_{n=0_i}^{a_i} \sum_{m=1}^{2n+1} (v_{0,\ldots,a})^{(i)\wedge}(n,m) \sum_{j=1}^{3} \sum_{p=0_j}^{a_j} \sum_{q=1}^{2p+1} u^{(j)\wedge}(p,q) \delta_{ij}\delta_{np}\delta_{mq} \\
&= \sum_{i=1}^{3} \sum_{n=0_i}^{a_i} \sum_{m=1}^{2n+1} (v_{0,\ldots,a})^{(i)\wedge}(n,m) \sum_{j=1}^{3} \sum_{p=0_j}^{a_j} \sum_{q=1}^{2p+1} u^{(j)\wedge}(p,q) \sum_{k=1}^{N} \mathrm{a}_k^{j,p,q} \cdot h_{n,m}^{(i),R}(x_k) \\
&= \sum_{j=1}^{3} \sum_{p=0_j}^{a_j} \sum_{q=1}^{2p+1} u^{(j)\wedge}(p,q) \sum_{k=1}^{N} \mathrm{a}_k^{j,p,q} \cdot v_{0,\ldots,a}(x_k).
\end{aligned}$$

□

A simple idea to reduce the total amount of weights is performed in the next corollary.

Corollary 5.2.2. *Under the assumption of Lemma 5.2.1, the formula*

$$(v_{0,\ldots,a}, u)_h = \sum_{k=1}^{N} \mathrm{a}_k \cdot v_{0,\ldots,a}(x_k)$$

holds for all weights $\mathrm{a}_1, \ldots, \mathrm{a}_N$ *satisfying the linear equations*

$$\sum_{k=1}^{N} \mathrm{a}_k \cdot h_{n,m}^{(i),R}(x_k) = u^{(i)\wedge}(n,m), \quad n = 0_i, \ldots, a_i; \;\; m = 1, \ldots, 2n+1.$$

Proof. We let $\mathrm{a}_k = \sum_{j=1}^{3} \sum_{p=0_j}^{a_j} \sum_{q=1}^{2p+1} u^{(j)\wedge}(p,q) \mathrm{a}_k^{j,p,q}$, $k = 1, \ldots, N$, and apply Lemma 5.2.1. □

Corollary 5.2.2 immediately leads to the following consequence.

Corollary 5.2.3. *If* $\mathrm{a}_1, \ldots, \mathrm{a}_N \in \mathbb{R}^3$ *are weights satisfying the linear equations*

$$\sum_{k=1}^{N} \mathrm{a}_k \cdot h_{n,m}^{(i),R}(x_k) = (u_{0,\ldots,a})^{(i)\wedge}(n,m), \tag{5.2.4}$$

$n \in \mathbb{N}_{0_i}$; $m = 1, \ldots, 2n+1$; $i = 1,2,3$, *then*

$$(v_{0,\ldots,a}, u_{0,\ldots,a})_h = \sum_{k=1}^{N} \mathrm{a}_k \cdot v_{0,\ldots,a}(x_k),$$

where we have used the canonical abbreviation

$$u_{0,\ldots,a} = \sum_{j=1}^{3} \sum_{p=0_j}^{a_j} \sum_{q=1}^{2p+1} u^{(i)\wedge}(p,q) h_{p,q}^{(j),R}.$$

Setting $u = h_{p,q}^{(j),R}$ we find:

Corollary 5.2.4. *If* $\mathrm{a}_1, \ldots, \mathrm{a}_N \in \mathbb{R}^3$ *are weights satisfying the linear equations*

$$\sum_{k=1}^{N} \mathrm{a}_k \cdot h_{n,m}^{(i),R}(x_k) = (h_{p,q}^{(j),R})^{(i)\wedge}(n,m) = (h_{p,q}^{(j),R}, h_{n,m}^{(i),R})_h = \delta_{ji}\delta_{pn}\delta_{qm}, \tag{5.2.5}$$

for $n \in \mathbb{N}_{0_i}$; $m = 1, \ldots, 2n+1$; $i = 1,2,3$, *then*

$$\begin{aligned}(v_{0,\ldots,a}, h_{p,q}^{(j),R})_h &= (v_{0,\ldots,a})^{(j)\wedge}(p,q) \\ &= \sum_{k=1}^{N} \mathrm{a}_k \cdot v_{0,\ldots,a}(x_k).\end{aligned}$$

Summarizing the results we are able to present a fully discrete Fourier approximation for a vector function $v_{0,\ldots,a} \in h_{0,\ldots,a}|_\Sigma$ in the following way:

Theorem 5.2.5. *Suppose that* $\{x_1, \ldots, x_N\} \subset \Sigma$, $3N \geq M$, $M = 3(a+1)^2$, $a > 1$, *is an admissible system of order* a *on* Σ *with respect to* $h_{0,\ldots,a}$. *Furthermore, assume that from a vector function* $v_{0,\ldots,a} \in h_{0,\ldots,a}$ *there are known the vector values* $v_{0,\ldots,a}(x_k) = v_k \in \mathbb{R}^3$, $k = 1, \ldots, N$. *Then*

$$v_{0,\ldots,a}(x) = \sum_{i=1}^{3} \sum_{n=0_i}^{a_i} \sum_{m=1}^{2n+1} \sum_{k=1}^{N} \mathrm{a}_k \cdot v_k h_{n,m}^{(i),R}(x), \quad x \in \Sigma, \tag{5.2.6}$$

provided that the weights $\mathfrak{a}_1, \ldots, \mathfrak{a}_N \in \mathbb{R}^3$ *satisfy the linear equations*

$$\sum_{k=1}^{N} \mathfrak{a}_k \cdot h_{p,q}^{(j),R}(x_k) = \delta_{ij}\delta_{pn}\delta_{qm} \tag{5.2.7}$$

for $i, j = 1, 2, 3$; $p = 0_j, \ldots, a_j$; $q = 1, \ldots, 2p+1$; $n = 0_i, \ldots, a_i$; $m = 1, \ldots, 2n+1$.

Proof. The Fourier expansion of $v_{0,\ldots,a}$ reads as follows

$$v_{0,\ldots,a}(x) = \sum_{i=1}^{3} \sum_{n=0_i}^{a_i} \sum_{m=1}^{2n+1} (v_{0,\ldots,a})^{(i)\wedge}(n,m) h_{n,m}^{(i),R}(x), \quad x \in \Sigma.$$

This can be rewritten in the form

$$v_{0,\ldots,a}(x) = \sum_{i=1}^{3} \sum_{n=0_i}^{a_i} \sum_{m=1}^{2n+1} \sum_{j=1}^{3} \sum_{p=0_j}^{a_j} \sum_{q=1}^{2p+1} \delta_{ij}\delta_{pn}\delta_{qm} (v_{0,\ldots,a})^{(j)\wedge}(p,q) h_{n,m}^{(i),R}(x).$$

Observing the linear equations (5.2.7) we find by resubstitution of the Fourier expansion

$$\begin{aligned} v_{0,\ldots,a}(x) &= \sum_{i=1}^{3} \sum_{n=0_i}^{a_i} \sum_{m=1}^{2n+1} \sum_{j=1}^{3} \sum_{p=0_j}^{a_j} \sum_{q=1}^{2p+1} \sum_{k=1}^{N} \mathfrak{a}_k \cdot h_{p,q}^{(j),R}(x_k) (v_{0,\ldots,a})^{(j)\wedge}(p,q) h_{n,m}^{(i),R}(x) \\ &= \sum_{i=1}^{3} \sum_{n=0_i}^{a_i} \sum_{m=1}^{2n+1} \sum_{k=1}^{N} \mathfrak{a}_k \cdot v_{0,\ldots,a}(x_k) h_{n,m}^{(i),R}(x). \end{aligned}$$

This is the desired result. □

The problem of representing a vector function $v_{0,\ldots,a} \in h_{0,\ldots,a}|_\Sigma$ from known vector values $v_{0,\ldots,a}(x_k) = v_k$, $k = 1, \ldots, N$, is the solution of the linear system (5.2.5), (5.2.7), respectively. Obviously

$$\sum_{k=1}^{N} \mathfrak{a}_k \cdot h_{n,m}^{(i),R}(x_k) = \delta_{ij}\delta_{pn}\delta_{qm}, \tag{5.2.8}$$

$n \in \mathbb{N}_{0_i}$; $m = 1, \ldots, 2n+1$; $i = 1, 2, 3$, is equivalent to

$$\begin{aligned} \sum_{i=1}^{3} \sum_{n=0_i}^{a_i} \sum_{m=1}^{2n+1} \left(\sum_{k=1}^{N} \mathfrak{a}_k \cdot h_{n,m}^{(i),R}(x_k) \right) h_{n,m}^{(i),R}(x) &= \sum_{i=1}^{3} \sum_{n=0_i}^{a_i} \sum_{m=1}^{2n+1} \delta_{ij}\delta_{pn}\delta_{qm} h_{n,m}^{(i),R}(x) \\ &= h_{p,q}^{(j),R}(x), \end{aligned} \tag{5.2.9}$$

$x \in \Sigma$, and

$$\sum_{k=1}^{N} a_k \cdot h_{n,m}^{(i),R}(x_k) = (u_{0,\ldots,a})^{(i)\wedge}(n,m), \tag{5.2.10}$$

for $n \in \mathbb{N}_{0_i}$; $m = 1, \ldots, 2n+1$; $i = 1,2,3$, is equivalent to

$$\sum_{i=1}^{3} \sum_{n=0_i}^{a_i} \sum_{m=1}^{2n+1} \left(\sum_{k=1}^{N} a_k \cdot h_{n,m}^{(i),R}(x_k) \right) h_{n,m}^{(i),R}(x) = u_{0,\ldots,a}(x), \quad x \in \Sigma. \tag{5.2.11}$$

Using the tensor notation the left hand side of (5.2.9) and (5.2.11) can be written as follows:

$$\begin{aligned} &\sum_{i=1}^{3} \sum_{n=0_i}^{a_i} \sum_{m=1}^{2n+1} \left(\sum_{k=1}^{N} a_k \cdot h_{n,m}^{(i),R}(x_k) \right) h_{n,m}^{(i),R}(x) \\ = &\sum_{k=1}^{N} \left(\sum_{i=1}^{3} \sum_{n=0_i}^{a_i} \sum_{m=1}^{2n+1} h_{n,m}^{(i),R}(x) \otimes h_{n,m}^{(i),R}(x_k) \right) a_k \\ = &\sum_{k=1}^{N} \mathbf{k}_{h_{0,\ldots,a}}(x, x_k) a_k. \end{aligned}$$

In other words, the coefficients $a_1, \ldots, a_N \in \mathbb{R}^3$ corresponding to (5.2.8) can be obtained by solving

$$\sum_{k=1}^{N} \mathbf{k}_{h_{0,\ldots,a}}(x_s, x_k) a_k = h_{p,q}^{(j),R}(x_s), \quad s = 1, \ldots, N. \tag{5.2.12}$$

Analogously, the coefficients $a_1, \ldots, a_N \in \mathbb{R}^3$ corresponding to (5.2.10) are solutions of the linear equations

$$\sum_{k=1}^{N} \mathbf{k}_{h_{0,\ldots,a}}(x_s, x_k) a_k = u_{0,\ldots,a}(x_s), \quad s = 1, \ldots, N. \tag{5.2.13}$$

Note that

$$\begin{pmatrix} h^{(1),R}_{0,1}(x_1)...h^{(1),R}_{a_1,2a_1+1}(x_1)...h^{(3),R}_{1,1}(x_1)...h^{(3),R}_{a_3,2a_3+1}(x_1) \\ \vdots \\ \vdots \\ h^{(1),R}_{0,1}(x_N)...h^{(1),R}_{a_1,2a_1+1}(x_N)...h^{(3),R}_{1,1}(x_N)...h^{(3),R}_{a_3,2a_3+1}(x_N) \end{pmatrix} \begin{pmatrix} (h^{(1),R}_{0,1}(x_1))^T & \dots & (h^{(1),R}_{0,1}(x_N))^T \\ \vdots & & \vdots \\ (h^{(1),R}_{a_1,2a_1+1}(x_1))^T & \dots & (h^{(1),R}_{a_1,2a_1+1}(x_N))^T \\ (h^{(2),R}_{1,1}(x_1))^T & \dots & (h^{(2),R}_{1,1}(x_N))^T \\ \vdots & & \vdots \\ (h^{(2),R}_{a_2,2a_2+1}(x_1))^T & \dots & (h^{(2),R}_{a_2,2a_2+1}(x_N))^T \\ (h^{(3),R}_{1,1}(x_1))^T & \dots & (h^{(3),R}_{1,1}(x_N))^T \\ \vdots & & \vdots \\ (h^{(3),R}_{a_3,2a_3+1}(x_1))^T & \dots & (h^{(3),R}_{a_3,2a_3+1}(x_N))^T \end{pmatrix}$$

$$= \begin{pmatrix} \mathbf{k}_{h_{0,\dots,a}}(x_1,x_1) & \dots & \mathbf{k}_{h_{0,\dots,a}}(x_1,x_N) \\ \vdots & & \vdots \\ \mathbf{k}_{h_{0,\dots,a}}(x_N,x_1) & \dots & \mathbf{k}_{h_{0,\dots,a}}(x_N,x_N) \end{pmatrix} \tag{5.2.14}$$

and the last $3N \times 3N$ block matrix is of rank M (since $\{x_1,\dots,x_N\} \subset \Sigma$ is an admissible system of order a on Σ with respect to $h_{0,\dots,a}$). Therefore, in case of $3N = M$, the linear systems (5.2.12), (5.2.13) are uniquely solvable.

5.3 Spline Representation of Vector Functions on Regular Surfaces

The interpolation problem for discussion can be formulated as follows. For an unknown $v \in h$, given the data points $(x_i, v(x_i)) \in \Sigma \times \mathbb{R}^3$, $i = 1,\dots,N$, find the smoothest vector field in the set I_N of all interpolants in h, namely

$$I_N = \{w \in h \,|\, w(x_l) = f(x_l),\ l = 1,\dots,N\}.$$

By the smoothest field, we mean the one for which the norm is minimized in h. In doing so we essentially follow the paper [13].

To deal with this problem, we discuss some preliminary results.

Lemma 5.3.1. *Let X_N be a system of N points $x_1,\dots,x_N$ on the surface Σ such that $x_n \neq x_m$ for $n \neq m$. Then the vector fields $\mathbf{k}(\cdot,x_l)\varepsilon^i$, $l = 1,\dots,N$, $i = 1,2,3$, are linearly independent.*

Proof. From the viewpoint of the theory of boundary value problems, $\mathbf{k}(\cdot,\cdot)$ is Green's matrix of the inner Dirichlet problem in vector potential theory for the sphere centered at the origin with radius R. Hence, if $x_n \neq x_m$ for $n \neq m$, then the linear independence of $\mathbf{k}(\cdot, x_l)\varepsilon^i$, $l = 1, \ldots, N$; $i = 1, 2, 3$, follows from the well-known arguments, e.g., to be found in [28]. □

From Lemma 5.3.1, another result follows immediately.

Lemma 5.3.2. *The* $3N \times 3N$ *block matrix*

$$\begin{pmatrix} \mathbf{k}(x_1, x_1) & \ldots & \mathbf{k}(x_1, x_N) \\ \vdots & & \vdots \\ \mathbf{k}(x_N, x_1) & \ldots & \mathbf{k}(x_N, x_N) \end{pmatrix} \tag{5.3.1}$$

is nonsingular, positive definite, and symmetric.

Proof. The matrix is a Gram matrix of linearly independent elements. □

In fact, the matrix (5.3.1) can be partitioned as it stands by conventional methods, such as the usual square-root method (Cholesky decomposition) or QR-decomposition, for which powerful computer routines are readily available (see [4] for example). On the other hand, the matrix (5.3.1) can be interpreted as a block positive-definite matrix. In this case, the block variant of the usual square-root method (see [1]) is known, too.

In the sequel, we use the following notation.

Let X_N be a nodal system on Σ, i.e., a system of distinct points $x_1, \ldots, x_N$ on Σ ($x_n \neq x_m$ for $n \neq m$). Then $\mathcal{S}_N = \mathcal{S}_N(x_1, \ldots, x_N)$ denotes the space of all vector fields $s \in h$ of the form

$$s(x) = \sum_{l=1}^{N} \mathbf{k}(x, x_l)\mathfrak{a}_l, \quad x \in \overline{\Sigma_{inf}^{ext}}, \tag{5.3.2}$$

where $\mathfrak{a}_1, \ldots, \mathfrak{a}_N \in \mathbb{R}^3$ are arbitrary coefficients. The kernel $\mathbf{k}$ can be decomposed as given by (5.1.7).

From Theorem 5.1.1 we now obtain the following result.

Lemma 5.3.3. *Let* s *be a vector field of class* $\mathcal{S}_N$ *of the form (5.3.2). Then, for each* $v \in h$,

$$(v, s)_h = \sum_{l=1}^{N} v(x_l)\mathfrak{a}_l.$$

Proof. It is easy to see that

$$\sum_{l=1}^{N}\sum_{i=1}^{3}(\mathrm{a}_l\cdot\varepsilon^i)\left(\mathbf{k}(\cdot,x_l)\varepsilon^i,\, v\right)_h=\sum_{l=1}^{N}v(x_l)\mathrm{a}_l.$$

Noting the identity (5.1.4), we derive the desired result. □

Lemma 5.3.4. *There exists a unique element $s\in\mathcal{S}_N\cap I_N$, which we denote in brief by s_N.*

Proof. Any $s\in h$ of the form (5.3.2) involves a total of N coefficients $\mathrm{a}_1,\ldots,\mathrm{a}_N\in\mathbb{R}^3$. Hence, $s\in I_N$ leads to N linear equations in these coefficients:

$$s(x_k)=\sum_{l=1}^{N}\mathbf{k}(x_k,x_l)\mathrm{a}_l,\quad k=1,\ldots,N.$$

This system is uniquely solvable provided that X_N is a nodal system as described above. □

From Lemma 5.3.3, we obtain by straightforward calculations the following result.

Lemma 5.3.5. *If $v\in I_N$, then*

$$\|v\|_h^2=\|s_N\|_h^2+\|s_N-v\|_h^2.$$

Summarizing the above results, we finally obtain the following theorem.

Theorem 5.3.6. *Let X_N be a nodal system on Σ. Then the interpolation problem*

$$\|s_N\|_h=\inf_{v\in I_N}\|v\|_h$$

is well posed in the sense that its solution exists, is unique, and depends continuously on the data $f(x_1),\ldots,f(x_N)$. The uniquely determined solution s_N is given in the explicit form

$$s_N(x)=\sum_{l=1}^{N}\mathbf{k}(x,x_l)\mathrm{a}_l,\quad x\in\overline{\Sigma_{inf}^{ext}},\tag{5.3.3}$$

where the coefficients $\mathrm{a}_1,\ldots,\mathrm{a}_N\in\mathbb{R}^3$ satisfy the linear equations

$$f(x_k)=\sum_{l=1}^{N}\mathbf{k}(x_k,x_l)\mathrm{a}_l,\quad k=1,\ldots,N.\tag{5.3.4}$$

For every nodal system X_N of N points $x_1, \ldots, x_N$ on Σ and for every vector field $v \in h$, there exists a unique element $s_{N;v} \in \mathcal{S}_N(x_1, \ldots, x_N)$ satisfying the conditions

$$v(x_l) = s_{N;v}(x_l), \quad l = 1, \ldots, N.$$

Let Θ_N denote the X_N−width on Σ, i.e., the maximal distance for any point of the surface Σ to the system X_N:

$$\Theta_N = \max_{x \in \Sigma} \left(\min_{y \in X_N} |x - y| \right).$$

Theorem 5.3.7. *Suppose that v is of class h. Let X_N be a nodal system on Σ. Then there exists a positive constant B (dependent on Σ, but independent of v) such that*

$$\sup_{x \in \Sigma} |v(x) - s_{N;v}(x)| \leq B \Theta_N \|v\|_h.$$

Proof. For any given $x \in \Sigma$, there exists a point $x_l \in X_N$ with $|x - x_l| \leq \Theta_N$. Now $v(x_l) = s_{N;v}(x_l)$, and thus it is clear that

$$s_{N;v}(x) - v(x) = (s_{N;v}(x) - s_{N;v}(x_l)) - (v(x) - v(x_l)).$$

Observing the reproducing property (5.1.4) and applying the Cauchy-Schwarz inequality, we find that

$$\begin{aligned} |v(x) - v(x_l)| &\leq (\beta(x, x_l))^{1/2} \|v\|_h, \\ |s_{N;v}(x) - s_{N;v}(x_l)| &\leq (\beta(x, x_l))^{1/2} \|s_{N;v}\|_h, \end{aligned} \tag{5.3.5}$$

where

$$\beta(x, x_l) = \sum_{i=1}^{3} \varepsilon^i \cdot (\mathbf{k}(x, x) - \mathbf{k}(x, x_l) - \mathbf{k}(x_l, x) + \mathbf{k}(x_l, x_l)) \, \varepsilon^i.$$

Now $s_{N;v}$ is the smoothest h interpolant, that is to say

$$\|s_{N;v}\|_h \leq \|v\|_h. \tag{5.3.6}$$

Consequently, from (5.3.5) and (5.3.6), the triangle inequality gives

$$|s_{N;v}(x) - v(x)| \leq 2(\beta(x, x_l))^{1/2} \|v\|_h.$$

From considerations given in [10] and [13], it follows that there exists a constant B (dependent on Σ, but independent of v) such that

$$|\beta(x, x_l)| \leq \frac{1}{4} B |x - x_l|^2.$$

Therefore we may deduce that

$$|s_{N;v}(x) - v(x)| \leq B \Theta_N \|v\|_h,$$

and this is the desired result. □

Summarizing our results, we therefore obtain the following theorem.

Theorem 5.3.8. *Suppose that v is a vector field of class h, and X_N is a nodal system of N points $x_1, \ldots, x_N$ on Σ. Let $s_{N;v}$ denote the uniquely determined solution of the minimization problem*

$$\|s_{N;v}\|_h = \inf_{w \in I_N^v} \|w\|_h,$$

where

$$I_N^v = \{w \in h \,|\, w(x_l) = v(x_l),\ l = 1, \ldots, N\}.$$

Then there exists a positive constant D (dependent on the geometry of Σ, but not on the field v) such that

$$\sup_{x \in \Sigma} |s_{N;v}(x) - v(x)| \leq D \Theta_N \|v\|_h.$$

Theorem 5.3.8 gives rise to the following conclusions. Let f be an element of the class $h|_\Sigma$ of all restrictions $v|_\Sigma$ of elements $v \in h$ to the surface Σ, and let $\{X_N\}$ be a sequence of nodal systems X_N such that $\Theta_N \to 0$ as $N \to \infty$. Then, for given $f \in h|_\Sigma$ the solution of the boundary value problem

$$v \in h|_{\overline{\Sigma^{ext}}}, \quad v|_\Sigma = f$$

can be arbitrarily well approximated in the sense that, for every $\varepsilon > 0$, there exists an integer $N = N(\varepsilon)$ and a linear combination

$$s_{N;v}(x) = \sum_{l=1}^{N} \sum_{i=1}^{3} \mathrm{a}_l^i \mathbf{k}(x, x_l) \varepsilon^i$$

uniquely given by

$$\sum_{l=1}^{N}\sum_{i=1}^{3} a_l^i \mathbf{k}(x, x_l)\varepsilon^i = f(x_l), \quad l = 1, \ldots, N,$$

such that

$$\sup_{x\in\Sigma} |s_{N;v}(x) - v(x)| \leq \varepsilon.$$

Thus we have developed a constructive method for solving the boundary value problem, provided that the boundary values f are elements of the class $h|_\Sigma$.

5.4 Theoretical Conclusions Concerning Spline Interpolation in h

In many problems (for example, the wind field determination on the topography), we are faced with the situation in which only a discrete set of vector data $(x_l, f(x_l))$, $l = 1, \ldots, N$, is available. To deal with this problem by applying our constructive method given in Section 5.3, we must assume that the values f are of class $h|_\Sigma$. From a theoretical point of view, however, it is of interest to discuss the situation in which the values f are assumed to be in $c(\Sigma)$, but not necessarily in $h|_\Sigma$. To consider this case, we should recall (see Theorem 4.2.5) that the vector outer harmonics restricted to Σ form a dense subset in $c(\Sigma)$ (in the sense of uniform topology on Σ). Hence $h|_\Sigma$, considered as a subset of $c(\Sigma)$ containing the space of vector outer harmonics restricted to Σ is dense in $c(\Sigma)$, too. Therefore, an extended version of Helly's theorem, due to [37], enables us to conclude that, for any positive number $\varepsilon > 0$, for any prescribed set $\tilde{X}_M$ of M points $\tilde{x}_1, \ldots, \tilde{x}_M$ on Σ, and for any element $f \in c(\Sigma)$, there exists an element g of the space $h|_\Sigma$ in the $\varepsilon-$neighbourhood of f such that $f(\tilde{x}_l) = g(\tilde{x}_l)$, $l = 1, \ldots, M$.
Combining these results, we therefore obtain the following theorem (see [10, 13]).

Theorem 5.4.1. *Let $\tilde{X}_M$ be a nodal set of M points $\tilde{x}_1, \ldots, \tilde{x}_M$ on Σ. Suppose that X_N is a sequence of nodal systems X_N on Σ such that $\tilde{X}_M \subset X_N$ for all N and $\Theta_N \to 0$ as $N \to \infty$. Then any $f \in c(\Sigma)$ can be arbitrarily well approximated in the sense that, for every $\varepsilon > 0$, there exists an integer $N = N(\varepsilon)$ and a linear combination*

$$s_N(x) = \sum_{l=1}^{N} \mathbf{k}(x, x_l) b_l, \quad x \in \Sigma, \tag{5.4.1}$$

such that

$$s_N(\tilde{x}_k) = f(\tilde{x}_k), \quad k = 1, \dots, M,$$

and

$$\sup_{x \in \Sigma} |s_N(x) - f(x)| \leq \varepsilon.$$

As in scalar potential theory [10], we are unable to find a suitable method for determining explicitly the linear combination (5.4.1) which realizes Theorem 5.4.1. In other words, the theoretical problem of approximating a continuous vector field f from discretely given data points on the surface Σ can be answered only in a nonconstructive way. Nevertheless, our theoretical result (Theorem 5.4.1) shows us that the kernel matrix (5.1.3) can be used to construct basis systems with continuous restrictions on Σ, i.e.,

$$c(\Sigma) = \overline{span_{x \in \mathcal{X}} \mathbf{k}(\cdot, x) \varepsilon^i|_\Sigma},$$

in the sense of the $c-$topology on Σ, provided that $\mathcal{X}$ is the union of a system of nodal systems X_N on Σ with $\Theta_N \to 0$ as $N \to \infty$.

5.5 Numerical Aspects of Vector Field Approximations

We provide some numerical examples for the approximation of vector fields on regular surfaces by using the techniques developed in Section 5.2. The described approach to model vector fields on regular surfaces considers the calculation of coefficients by solving the linear equation system (5.2.7). The calculated coefficients are then used to approximate a vector function on Σ by using function values given on the regular surface Σ (see (5.2.6)).
The topography data is taken from the GLOBE Project of NOAA's (National Oceanic and Atmospheric Administration) National Geophysical Data Center.
Let us now face the topography of the area of Rheinland-Pfalz (Palatinate), where the Forest Research Institute Rheinland-Pfalz ("Forschungsanstalt für Waldökologie und Forstwirtschaft (FAWF) in Rheinland-Pfalz") records quantities in fields of temperature, humidness, precipitation and wind velocity. In a cooperation between the FAWF and the Geomathematics Group Kaiserslautern models [15] based on scalar interpolation are already developed for temperature, humidness and precipitation. These observational quantities are acquired in Rheinland-Pfalz at 15 stations (see left illustration in Figure 5.5.1). Numerical results are already obtained by spherical spline interpolation (as given in Figure 5.5.1, right hand side) of the absolute value (norm) and the direction of the wind field in [16]. We are interested in the meso-scale wind propagation. The meso-scale specifies, as e.g., described in [36], the dimension of spatially-horizontal phenomena regarding their resolution. The spatial resolution ranges between 2 and 2000 kilometers. Thus, an approximation with low degrees is sufficient. In the following we will exemplarily show how the approximation of wind field data can be performed by the use of spline interpolation with vector outer harmonics. The wind measurements are taken at points of Rheinland-Pfalz at different heights but always having a fixed distance of 10 meters to the topography. Figure 5.5.2 shows the values for 22.01.2002 (at 4.p.m.). This set of values will be used in the following to demonstrate exemplarily the interpolation technique for continuous vector functions on a regular surface. In this case the topography of Rheinland-Pfalz serves as the regular surface. Regarding [34] the wind field is dominated by horizontal vector components, instead of three dimensional currents. The wind measurements consist out of a an angle which gives the direction and the absolute value, therefore, our calculations are also based on horizontal vector values.

First we want to validate the interpolation behavior by taking a field with constant directions and values into account and observe the interpolation result. We choose 9 input vectors $f_1, \ldots, f_9$ at the corresponding data points $x_1, \ldots, x_9$ (see Figure 5.5.3) and set in the first case all to the same direction (200°) and absolute values ($f_i = 1$ for $i = 1, \ldots, 9$). For the second example we take the direction from the wind measurements and let the values still be fix. And in the last case we take the measured wind speed and let the direction fix (200°).

The points $x_1, \ldots, x_9$ are from the area $5.9° - 8.9°$ East and $49° - 51°$ North. For the evaluation we use a grid of dimension 49×72.

We use (5.2.7), respectively (5.2.12), to retrieve the coefficients $\mathfrak{a}_k$, $k = 1, \ldots, 9$. The linear equation systems can be uniquely solved because $3N = M$, with $M = 27$. Therefore we solve the linear system

$$\mathbf{h} v_{\mathfrak{a}} = d, \quad \mathbf{h} \in \mathbb{R}^{M \times 3N}, \quad M = 3N,$$

where $\mathbf{h}$ is a matrix as given in (5.2.2), $v_{\mathfrak{a}}$ is a vector of dimension $3N$ representing the coefficients $\mathfrak{a}_k$, $k = 1, \ldots, 9$, i.e.,

$$v_{\mathfrak{a}} = \begin{pmatrix} \mathfrak{a}_1 \\ \mathfrak{a}_2 \\ \vdots \\ \mathfrak{a}_9 \end{pmatrix}$$

and d is a vector of dimension M with entries equal to 1 and 0.

We solve this linear equation system by standard routines provided by Matlab. In Figure 5.5.4 the result of the interpolation with degree $n = 2$ for the first case (direction and values are fix) is given. We obtain an exact interpolation with vector values equal to 1 and the direction being 200° at all points of the evaluation area. In Figure 5.5.5 and Figure 5.5.6 the results for the interpolation with degree $n = 2$ for the second case (direction from measurements and fix values) and the third case (fix direction and measured values) can be seen. We see that the strength of the interpolation procedure lies in the interpolation of the direction. For the absolute values the deviation in case the values are fix is high.

Now let us take into account the wind measurements of the 9 data points $x_1, \ldots, x_9$ (see Figure 5.5.3). Same calculations as above yield the interpolation result given in Figure 5.5.7. We observe that we obtain an interpolation with some irregularities in the northern river deltas (which can be seen in the bottom figure).

The next step considers all 15 measurement stations. What follows is that for degree $n = 2$ the dimension of the matrix $\mathbf{h}$ changes from (27×27) to (27×45), because $3N = 45$ and $M = 27$. That means that we are confronted with an under-determined linear equation system:

$$\mathbf{h}v_{\mathrm{a}} = d, \quad \mathbf{h} \in \mathbb{R}^{M\times 3N}, \quad M \leq 3N.$$

An under-determined set of equations usually has infinitely many solutions. Therefore we can replace the problem with a least-norm problem, i.e., looking for the solution with the smallest norm (see [23]).

Theorem 5.5.1. *If $\boldsymbol{h} \in \mathbb{R}^{M\times 3N}$ has rank M, then the solution of the least-norm problem of finding an $\tilde{v}_{\mathrm{a}}$ that satisfies*

$$\boldsymbol{h}\tilde{v}_{\mathrm{a}} = d$$

and

$$\|\tilde{v}_{\mathrm{a}}\|_2^2 \leq \|v_{\mathrm{a}}\|_2^2$$

for all v_{a} that satisfy $\boldsymbol{h}v_{\mathrm{a}} = d$ is unique and given by

$$\tilde{v}_{\mathrm{a}} = \boldsymbol{h}^T(\boldsymbol{h}\boldsymbol{h}^T)^{-1}d. \tag{5.5.1}$$

Proof. We first show that $\mathbf{h}\mathbf{h}^T$ is positive definite. We have

$$v_{\mathrm{a}}^T\mathbf{h}\mathbf{h}^T v_{\mathrm{a}} = (\mathbf{h}^T v_{\mathrm{a}})^T(\mathbf{h}^T v_{\mathrm{a}}) = \|\mathbf{h}^T v_{\mathrm{a}}\|_2 \geq 0$$

for all v_{a}. Moreover if $rank\mathbf{h} = M$, then $\|\mathbf{h}^T v_{\mathrm{a}}\|_2 = 0$ only if $v_{\mathrm{a}} = 0$. This means that $\mathbf{h}\mathbf{h}^T$ is positive definite, hence non singular. Next, we verify that $\tilde{v}_{\mathrm{a}}$ satisfies $\mathbf{h}\tilde{v}_{\mathrm{a}} = d$:

$$\mathbf{h}\tilde{v}_{\mathrm{a}} = (\mathbf{h}\mathbf{h}^T)(\mathbf{h}\mathbf{h}^T)^{-1}d = d.$$

Finally, we have to show that any other solution of the equations has a norm greater than $\|\tilde{v}_{\mathrm{a}}\|_2$. Suppose v_{a} satisfies $\mathbf{h}v_{\mathrm{a}} = d$. We have

$$\|v_{\mathrm{a}}^2\|_2 = \|\tilde{v}_{\mathrm{a}} + (v_{\mathrm{a}} - \tilde{v}_{\mathrm{a}})\|_2^2 = \|\tilde{v}_{\mathrm{a}}\|_2^2 + \|v_{\mathrm{a}} - \tilde{v}_{\mathrm{a}}\|_2^2 + 2\tilde{v}_{\mathrm{a}}^2(v_{\mathrm{a}} - \tilde{v}_{\mathrm{a}}).$$

Now, observe that

$$\begin{aligned}\tilde{v}_{\rm a}^2(v_{\rm a}-\tilde{v}_{\rm a}) &= (\mathbf{h}^T(\mathbf{h}\mathbf{h}^T)^{-1}d)^T(v_{\rm a}-\tilde{v}_{\rm a})\\ &= d^T(\mathbf{h}\mathbf{h}^T)^{-1}\mathbf{h}(v_{\rm a}-\tilde{v}_{\rm a})\\ &= 0\end{aligned}$$

because $\mathbf{h}v_{\rm a} = \mathbf{h}\tilde{v}_{\rm a} = d$. We therefore have

$$\|v_{\rm a}\|_2^2 = \|v_{\rm a}-\tilde{v}_{\rm a}\|_2^2 + \|\tilde{v}_{\rm a}\|_2^2 \geq \|\tilde{v}_{\rm a}\|_2$$

with equality only if $v_{\rm a} = \tilde{v}_{\rm a}$. In conclusion, if $\mathbf{h}v_{\rm a} = d$ and $v_{\rm a} \neq \tilde{v}_{\rm a}$ then

$$\|v_{\rm a}\|_2^2 > \|\tilde{v}_{\rm a}\|_2^2.$$

This proofs that $\tilde{v}_{\rm a}$ is the unique solution of the least-norm problem. □

This result means that we can solve the least-norm problem by solving

$$(\mathbf{h}\mathbf{h}^T)b = d \tag{5.5.2}$$

and then calculating $\tilde{v}_{\rm a} = \mathbf{h}^T b$. The equations (5.5.1) are a set of M linear equations in M variables, and are called the normal equations associated with the least-norm problem. The normal equations (5.5.2) can be solved by standard procedures using the Cholesky factorization of $\mathbf{h}^T\mathbf{h}$ or the QR factorization of $\mathbf{h}^T$.

In Figure 5.5.2 the set of 15 input points is given. The least-norm solution to that problem is given in Figure 5.5.8. We observe that we gain a smooth vector field approximation.

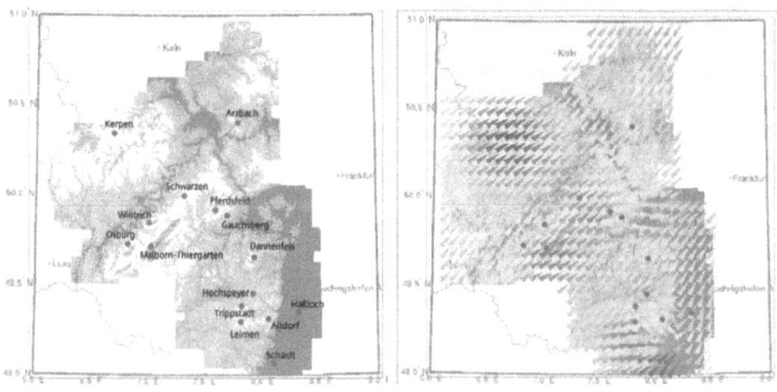

Figure 5.5.1: Left: Observational network of FAWF for wind measurements. Right: Scalar spline interpolation of wind field over Palatinate on 22.01.2002 (4pm). The values range between 1 and 5.7 m/s wind speed.

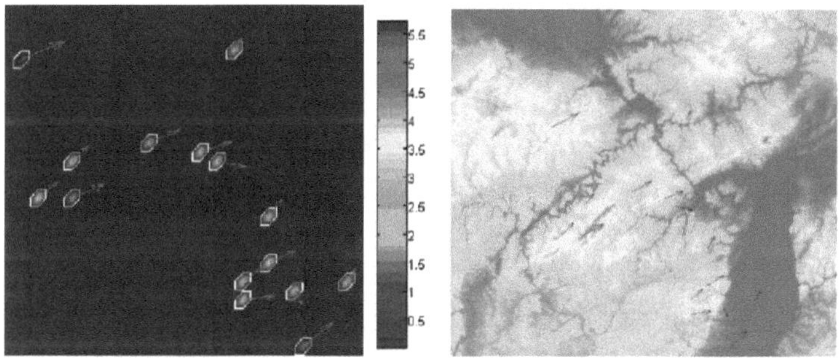

Figure 5.5.2: Input points from FAWF for the date 22.01.2002 at 4pm. Left: absolute value of the input points denoting the speed in m/s. Right: vectors on the topography of Rheinland-Pfalz ($5.9° - 8.9°$ East and $49° - 51°$ North).

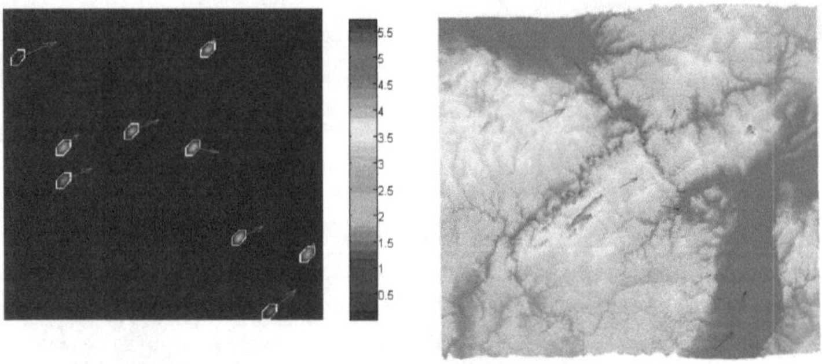

Figure 5.5.3: Set of 9 vector input points from the wind field measurements over Palatinate. Left: vectors versus their absolute value. Right: vectors versus the topography.

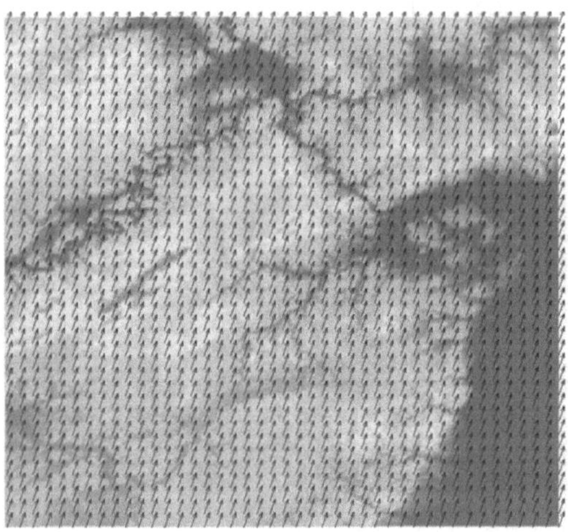

Figure 5.5.4: Interpolation (degree 2) of a vector field over the topography of Palatinate with fix direction (200°) and fix values $f_i = 1$, $i = 1, \dots, 9$ (marked in red). Plotted are the vectors versus the topography. The interpolation vectors have all the absolute value 1.

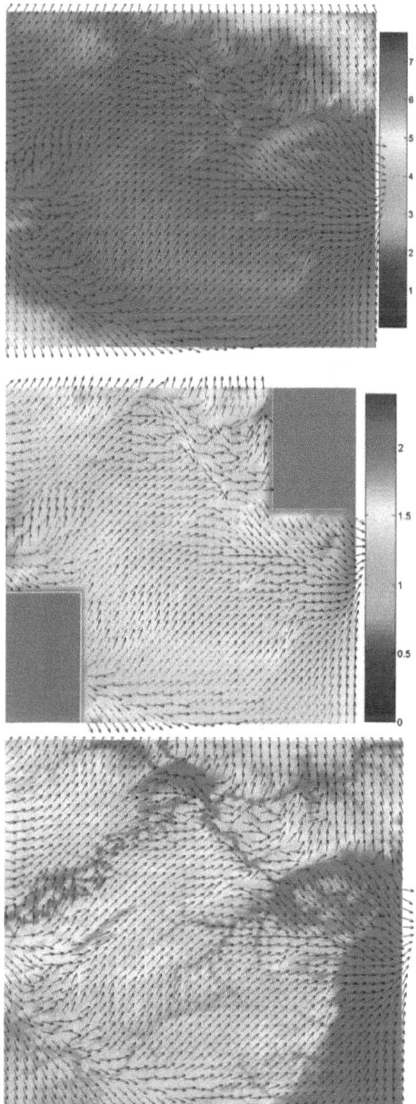

Figure 5.5.5: Interpolation (degree 2) of a vector field over the topography of Palatinate with measured direction and fix values $f_i = 1$, $i = 1, \ldots, 9$ (marked in red). Form top to bottom: Whole area with vectors plotted versus their absolute value. Boundaries are removed for better resolution of the absolute vector values. Vectors plotted versus the topography.

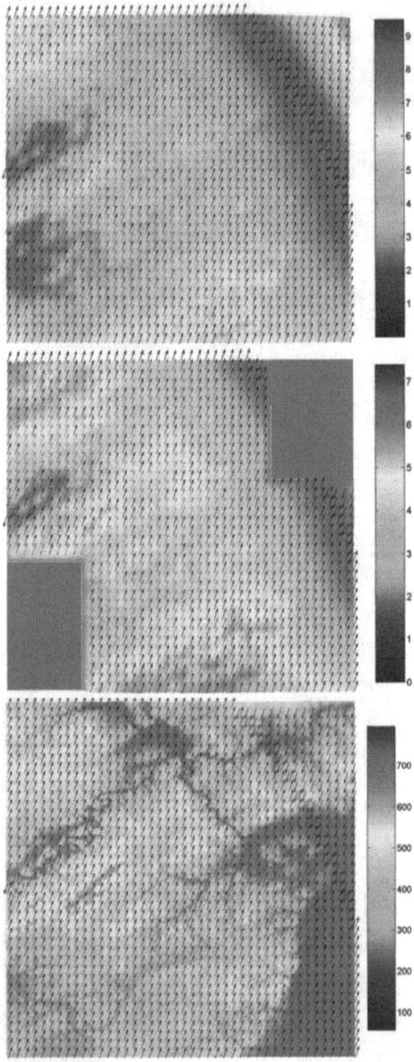

Figure 5.5.6: Interpolation (degree 2) of a vector field over the topography of Palatinate with fix direction and measured values $f_i = 1$, $i = 1, \dots, 9$ (marked in red). Form top to bottom: Whole area with vectors plotted versus their absolute value. Boundaries are removed for better resolution of the absolute vector values. Vectors plotted versus the topography.

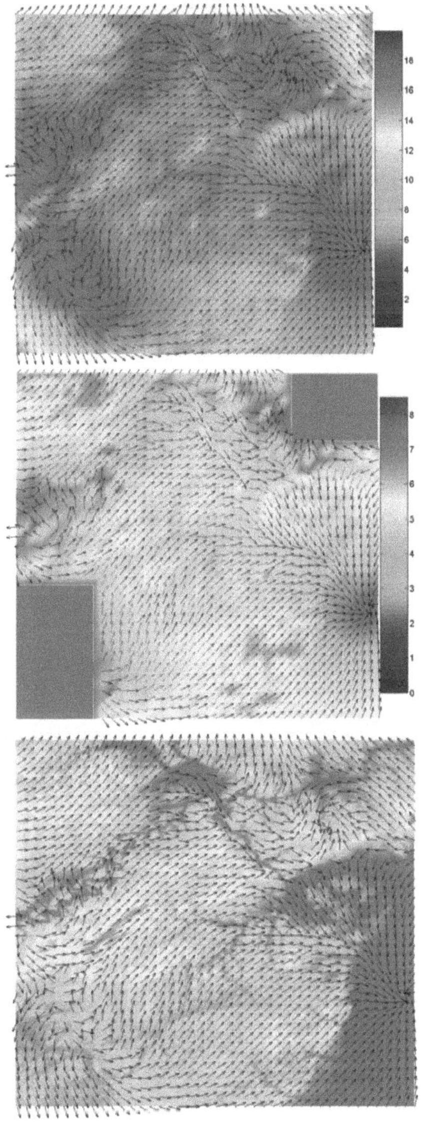

Figure 5.5.7: Interpolation (degree 2) with 9 vector input points (marked in red) from the wind field measurements over Palatinate. From top to bottom: Vectors versus their absolute value. Boundaries are removed for better resolution of the absolute vector values. Vectors plotted versus the topography.

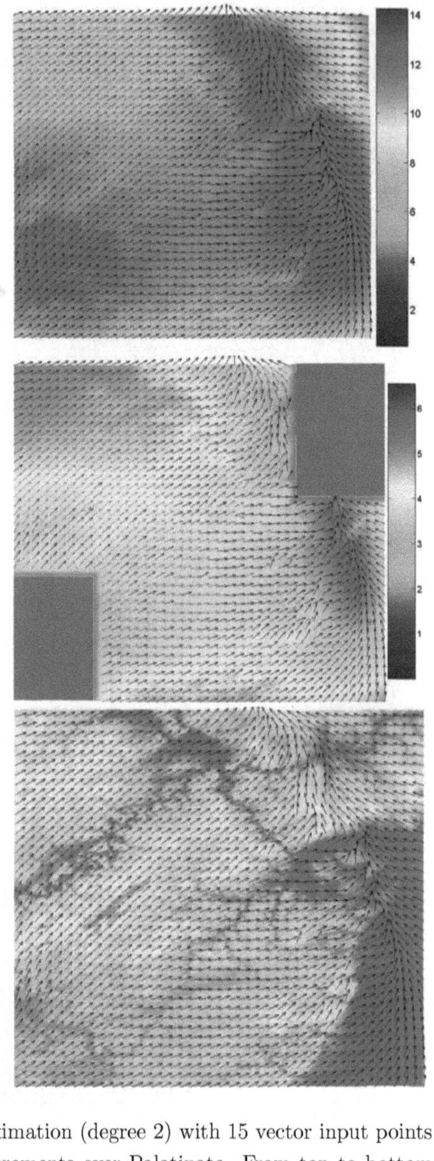

Figure 5.5.8: Approximation (degree 2) with 15 vector input points (marked in red) from the wind field measurements over Palatinate. From top to bottom: Vectors versus their absolute value. Boundaries are removed for better resolution of the absolute vector values. Vectors plotted versus the topography.

Chapter 6

Summary and Outlook

In this thesis we achieved two main goals. First (Section 4.3), we presented algorithms for the exact construction of scalar and vector spherical harmonics and their extension to scalar and vector outer harmonics. The basis for that is given by the exact calculation of homogeneous harmonic polynomials. One advantage of the exact generation of vector outer harmonics in terms of cartesian coordinates is that it avoids problems arising from singularities of a spherical coordinate system.
In a second step, we developed, based on vector outer harmonics, an approximate integration formulae, where the nodal points are not taken on the reference area of integration. The underlying concept extends the existing developments for scalar problems (as described in [18, 21]) by the representation for vector fields. We developed a Fourier variant for vector functions, where only values on a regular surface, e.g., the Earth topography, are given. The Fourier representation consists of coefficients which can be determined by solving a set of linear equations involving vector outer harmonics. At this point the generation (as presented in Section 4.3) of vector outer harmonics plays an important role.
Further, with tensor kernels composed from vector outer harmonics a spline interpolation procedure for continuous vector functions on regular surfaces is developed.

The methods developed in this thesis are a first step towards the approximation and interpolation of continuous vector functions on regular surfaces. Further investigation could be, e.g., done to develop the explicit representation for tensor kernels of vector outer harmonics. In the case of a more dense data situation concepts for multiscale approximation and wavelet theory can be continued straightforward from the developed settings of Chapter 5, hence, the exact computation of the vector basis functions (Section 4.4) can then be used to explicitly derive scaling functions.

To handle the numerical efforts when handling huge amounts of discretely given data one could extend the methods presented in [25]. There a domain decomposition method based on the Schwarz alternating algorithm for large symmetric positive definite matrices is presented. This algorithm enables the solution of extremely large equation systems from, e.g., spline interpolation problems. The procedure comprises first the split of the large linear equation systems into several smaller subsystems and then solves them alternating in an iterative algorithm.

Bibliography

[1] I.S. Beresin and N.P. Shidkow. *Numerische Methoden 2*. Berlin: Deutscher Verlag der Wissenschaften, 1971.

[2] J.M. Brudowsky. *Über die elementaren Darstellungen Cauchy Navierscher Splines*. Diploma Thesis, University of Kaiserslautern, Geomathematics Group, 1992.

[3] P.J. Davis. *Interpolation and Approximation*. Blaisdell Publishing Company, 1963.

[4] J.J. Dongarra, J.R. Bunch, C.B. Moler, and G.W. Stewart. *LINPACK Users Guide*. Philadelphia, PA:SIAM, 1979.

[5] A. R. Edmonds. *Angular Momentum in Quantum Mechanics*. Princeton University Press, 1957.

[6] M.J. Fengler. *Vector Spherical Harmonic and Vector Wavelet Based Non-Linear Galerkin Schemes for Solving the Incompressible Navier-Stokes Equation on the Sphere*. PhD Thesis, Geomathematics Group, University of Kaiserslautern, 2005.

[7] W. Freeden. Über eine Klasse von Integralformeln der Mathematischen Geodäsie. *Veröff. Geod. Inst. RWTH Aachen*, 27, 1979.

[8] W. Freeden. On the Approximation of External Gravitational Potential with Closed Systems of (Trial) Functions. *Bull. Géod.*, (54):1–20, 1980.

[9] W. Freeden. *Computation of Spherical Harmonics and Approximation by Spherical Harmonic Expansions*. The Ohio State University, Department of Geodetic Science and Surveying, Columbus, Ohio, OSU Report, 362, 1985.

[10] W. Freeden. A Spline Interpolation Method for Solving Boundary Value Problems of Potential Theory from Discretely Given Data. *Numer. Meth. Part. Diff. Equations*, (3):375–398, 1987.

[11] W. Freeden. *Multiscale Modelling of Spaceborn Geodata.* B.G. Teubner Verlag, 1999.

[12] W. Freeden and T. Gervens. Vector Spherical Spline Interpolation (Basic Theory and Computational Aspects). *Math. Meth. in the Appl. Sci.*, (16):151–183, 1991.

[13] W. Freeden, T. Gervens, and T. Mason, J.C.. A Minimum Norm Interpolation Method for Determining the Displacement Field of a Homogeneous Isotropic Elastic Body from Discrete Data. *IMA J. App. Math.*, (44):55–76, 1990.

[14] W. Freeden, T. Gervens, and M. Schreiner. *Constructive Approximation on the Sphere (With Applications to Geomathematics).* Oxford Science Publications, Clarendon, 1998.

[15] W. Freeden, O. Glockner, J. Rang, and G. Schüler. Von forstmeteorologischen Punktmessungen zu räumlich aggregierten Daten. *Forstw. Cbl.*, (119):332–349, 2000.

[16] W. Freeden, S. Gramsch, A. Luther, and G. Schüler. Konstruktive Approximation des Windfeldes - Entscheidungsgrundlage für waldbauliche Planungen. *Forstarchiv*, (76):44–53, 2005.

[17] W. Freeden and K. Hesse. Spline Modelling of Geostrophic Flow: Theoretical und Numerical Aspects. *Schriften zur Funktionalanalysis und Geomathematik*, 15, 2004.

[18] W. Freeden and V. Michel. *Multiscale Potential Theory (with Applications to Geoscience).* Birkhäuser, Basel, 2003.

[19] W. Freeden and R Reuter. Exact Computation of Spherical harmonics. *Computing*, (32):365–378, 1984.

[20] W. Freeden and R. Reuter. An Efficient Algorithm for the Generation of Homogeneous Harmonic Polynomials. *Scientific Software Systems (J.C. Mason, MG. Cox eds.), Chapman and Hall, 166-180*, 1990.

[21] W. Freeden and F. Schneider. Wavelet Approximation on Closed Surfaces and Their Application to Boundary-Value Problems of Potential Theory. *Math. Meth. in the Appl. Sci.*, (21):129–163, 1998.

[22] W. Freeden and U. Windheuser. Spherical Wavelet Transform and its Discretization. *Adv. Comput. Math.*, (5):51–94, 1996.

[23] Philip E. Gill, Walter Murray, and Margaret H. Wright. *Numerical Linear Algebra and Optimization.* Addison-Wesley Publishing Company, Redwood City, CA, Volume 1, 1991.

[24] M.E. Gurtin. *The Linear Theory of Elasticity.* Number VI a/2. Springer, 1972.

[25] K. Hesse. *Domain Decomposition Methods in Multiscale Determination from SST and SGG.* PhD Thesis, Geomathematics Group, University of Kaiserslautern, Shaker Verlag, 2002.

[26] E.W. Hobson. *The Theory of Spherical and Ellipsoidal Harmonics.* Reprint Chelsea Publishing Company, 1955.

[27] O.D. Kellogg. *Foundations of Potential Theory.* Frederick Ungar Publishing Company, 1929.

[28] V.D. Kupradze. Potential Methods in the Theory of Elasticity. *Jerusalem: Israel Program for Scientific Translations*, 1965.

[29] R. Litzenberger. *Pyramid Schemes for Harmonic Wavelets in Boundary-Value Problems.* PhD Thesis, Geomathematics Group, University of Kaiserslautern, Shaker Verlag, 2002.

[30] T. Maier. *Multiscale Geomagnetic Field Modeling from Satellite Data; Theoretical Aspects and Numerical Applications.* PhD Thesis, Geomathematics Group, University of Kaiserslautern, 2003.

[31] C. Müller. *Spherical Harmonics.* Lecture Notes in Mathematics, 17, Springer, 1966.

[32] C. Müller. *Foundations of the Mathematical Theory of Electromagnetic Waves.* Springer, 1969.

[33] H. Nutz. *A Unified Setup of Gravitational Field Observables.* PhD Thesis, Geomathematics Group, University of Kaiserslautern, Shaker Verlag, 2002.

[34] REKLIP. Klimaatlas Oberrhein Mitte-Süd. *VDF (ETH Zürich)*, 1996.

[35] C. Runge. Zur Theorie der eindeutigen analytischen Funktionen. *Acta. Math.*, (6):229–234, 1885.

[36] C. D. Schönwiese. *Klimatologie.* Eugen Ulmer Stuttgart, 1994.

[37] H. Yamabe. An Extension of Helly's Theorem. *Osaka Math. J.*, (2):15–17, 1950.

FSC
www.fsc.org
MIX
Papier | Fördert gute Waldnutzung
FSC® C083411

Zeitfracht Medien GmbH
Ferdinand-Jühlke-Straße 7
99095 Erfurt, Deutschland
produktsicherheit@kolibri360.de

Druck:
CPI Druckdienstleistungen GmbH
im Auftrag der
Zeitfracht Medien GmbH
Ein Unternehmen der Zeitfracht - Gruppe
Ferdinand-Jühlke-Str. 7
99095 Erfurt